SWEET DEAL,
BITTER LANDSCAPE

A volume in the series

Cornell Series on Land: New Perspectives on Territory, Development, and Environment

Edited by Wendy Wolford, Nancy Lee Peluso, and Michael Goldman

A list of titles in this series is available at cornellpress.cornell.edu.

SWEET DEAL, BITTER LANDSCAPE

Gender Politics and Liminality in Tanzania's New Enclosures

Youjin B. Chung

CORNELL UNIVERSITY PRESS   ITHACA AND LONDON

Copyright © 2023 by Cornell University

All rights reserved. Except for brief quotations in a review, this book, or parts thereof, must not be reproduced in any form without permission in writing from the publisher. For information, address Cornell University Press, Sage House, 512 East State Street, Ithaca, New York 14850. Visit our website at cornellpress.cornell.edu.

First published 2023 by Cornell University Press

Library of Congress Cataloging-in-Publication Data

Names: Chung, Youjin B., 1987– author.
Title: Sweet deal, bitter landscape : gender politics and liminality in Tanzania's new enclosures / Youjin B. Chung.
Description: Ithaca [New York] : Cornell University Press, 2023. | Series: Cornell series on land: new perspectives on territory, development, and environment | Includes bibliographical references and index.
Identifiers: LCCN 2023017732 (print) | LCCN 2023017733 (ebook) | ISBN 9781501772009 (hardcover) | ISBN 9781501772016 (paperback) | ISBN 9781501772030 (epub) | ISBN 9781501772023 (pdf)
Subjects: LCSH: EcoEnergy (Firm)—Influence. | Rural development—Tanzania. | Women—Tanzania—Social conditions. | Rural development—Political aspects—Tanzania. | Land tenure—Social aspects—Tanzania.
Classification: LCC HN797.Z9 C624 2023 (print) | LCC HN797.Z9 (ebook) | DDC 307.1/41209678—dc23/eng/20230712
LC record available at https://lccn.loc.gov/2023017732
LC ebook record available at https://lccn.loc.gov/2023017733

For umma and appa

Contents

Acknowledgments	ix
Abbreviations	xiii
Note on Currency	xv
Introduction	1
1. The Making of a Sweet Deal	27
2. The Making of a Bitter Landscape	53
3. On Being Counted: Gender, Property, and "the Family"	90
4. Governing Liminality: The Bio-necropolitics of Gender	105
5. Negotiating Liminality: Everyday Resistance and the Moral Economies of Difference	123
6. Of Privilege, Lawfare, and Perverse Resistance	149
Conclusion	169
Glossary of Swahili Terms	187
Notes	189
Bibliography	217
Index	243

Acknowledgments

This book would not have been possible without the trust, kindness, and generosity of many individuals and families in Tanzania who welcomed me into their lives. There are not enough words to describe my gratitude to them. In Bagamoyo, I had the good fortune of working with Francis Shang'a, whose formal title was research assistant but who did so much more, serving as a friend, interlocutor, and confidant with whom I shared both the challenging and rewarding moments of fieldwork. Christine Noe was a supportive and trusted mentor, and I thank her for facilitating my research affiliation with the Department of Geography at the University of Dar es Salaam. I am grateful for the friendship of Marjorie Mbilinyi, whose feminist scholarship and activism I admire deeply; Marge and the late mzee Simon took me in as family in Dar es Salaam and visited me in Bagamoyo, providing me with a sense of home away from home. Lussaga Kironde at Ardhi University graciously shared his expertise on land issues, particularly involuntary resettlement and compensation, and introduced me to key informants in the government in the early days of my research. I also benefited from exchanges with other scholars investigating land deal politics in Tanzania, including Chambi Chachage, Linda Engström, Richard Mbunda, Sina Schlimmer, and Emmanuel Sulle. I also thank the following individuals and activists for offering their wisdom and assistance at various points in my research: Stephen Chiombola, Gidufana Gafufen, Wilhelm Gidabuday, Lembulung M. Ole Kosyando, Magareth Maina, Sabrina Masinjila, Godfrey Massay, Samwel Mesiak, Elias Mtinda, Mary Ndaro, Valentin Ngorisa Olyang'iri, Ester Rwela, and Naomi Shadrack. I also thank the Tanzania Commission for Science and Technology, the Tanzania National Parks Authority, and local and regional government authorities for granting permission to conduct this research, as well as the Tanzania National Archives staff in Dar es Salaam for their patient assistance.

At Cornell University, where research for this book began, I had the special privilege of working with Wendy Wolford. Her commitment to agrarian justice and insistence on theoretical and methodological rigor continue to inspire me and my approach to research, teaching, and advising. Phil McMichael, Rachel Bezner Kerr, and Rick Schroeder were invaluable mentors; their sustained encouragement and critical observations pushed me to refine my arguments and better articulate the broader significance of my work. Each one of them and their scholarship influenced my thinking in more ways than I can imagine, and I feel so blessed to have had the opportunity to learn from them. I also valued conversations with Garry Thomas at Ithaca College. He was a tireless advocate of my work and generously donated his Tanzania library collections to me in retirement, a gift that I hope to pass on to my students someday.

Many friends and colleagues have read and commented on previous iterations of my writing on this book. I wish to acknowledge Ellie Andrews, Penelope Anthias, Alice Beban, Holly Buck, Andrew Curley, Ross Doll, Linda Engström, Hilary Faxon, Marie Gagné, Fernando Galeana-Rodriguez, Prabhat Gautham, Chuck Geisler, Asher Ghertner, Ritwick Ghosh, Tim Gorman, Maron Greenleaf, Elena Guzman, Kyle Harvey, Emily Hong, Veronica Jacome, Pauline Limbu, Christian Lund, Marjorie Mbilinyi, Matt Minarchek, Rebakah Daro Minarchek, Ryan Nehring, Gustavo Oliveira, Kasia Paprocki, Karla Peña, Tess Pendergrast, Katie Rainwater, Mattias Borg Rasmussen, Ewan Robison, Joeva Rock, Josh Savala, Annie Shattuck, Emmanuel Sulle, Tirza van Bruggen, Marygold Walsh-Dilley, and Michael Watts. Not everything they have read made it onto the pages of this book, but I am nonetheless grateful for their time, attention, and feedback that advanced my thinking.

I was fortunate to have Asha Best, Meg Mills-Novoa, and Sunaura Taylor as my fellow junior faculty colleagues at Clark University and the University of California at Berkeley. They offered immeasurable moral support as we juggled the demands of our positions and navigated the seemingly daunting process of writing a book. Alice Beban, Andrew Curley, Madeleine Fairbairn, Michael Mascarenhas, James McCarthy, Kasia Paprocki, Nancy Lee Peluso, Sara Smith, and Wendy Wolford provided important advice at different stages of book publishing. I thank my senior colleagues at the Energy and Resources Group and the Department of Environmental Science, Policy, and Management at UC Berkeley, especially my departmental mentors, Isha Ray and Rachel Morello-Frosch, for their wisdom and encouragement.

Across the Berkeley campus, Leo Arriola and Martha Saavedra warmly welcomed me to the African studies community and invited me to present a chapter at the UC-wide interdisciplinary Africanist Workshop. I thank them and other workshop participants, especially discussant Derick Fay, for their thoughtful engagement and constructive comments. The graduate students who enrolled in my Agrarian Questions seminar in the spring of 2020 and fall of 2021 also deserve recognition. Their reflections and lively discussions, particularly on agency, moral economy, and resistance, motivated the writing and rewriting of chapter 5. I also had the opportunity to present draft chapters at the George Perkins Marsh Institute at Clark University, the Department of Geography at UC Berkeley, and at the annual meetings of the African Studies Association and the American Association of Geographers. Much appreciation to Joanny

Bélair, Aharon de Grassi, Ashley Fent, Marie Gagné, Alicia Lazzarini, and Sina Schlimmer for inspiring generative conversations about global land investments in Africa at these conferences.

Funding for this research came from the Social Science Research Council's International Dissertation Research Fellowship and the Dissertation Proposal Development Fellowship. Various institutions and programs at Cornell University also provided financial support for research and writing: the Institute for the Social Sciences (Theme Project on Contested Global Landscapes), the Institute for African Development, the Office of International Programs at the College of Agriculture and Life Sciences (Initiative on Advancing Women in Agriculture through Research and Education), the Mario Einaudi Center for International Studies, the Department of Development Sociology, the Graduate School, the Center for the Study of Inequality, the Polson Institute for Global Development, the Society for the Humanities, and the Institute for Comparative Modernities. Clark University and UC Berkeley also provided funding for additional data collection, analysis, writing, and editing. Open access publishing was made possible by the generous support from the Berkeley Research Impact Initiative sponsored by the UC Berkeley Library, the Berkeley Excellence Accounts for Research Program, and the Energy and Resources Group.

Parts of chapter 2 appeared as "The Grass Beneath: Conservation, Agro-industrialization, and Land-Water Enclosures in Postcolonial Tanzania," *Annals of the American Association of Geographers* 109, no. 1 (2018):1–17, copyright © 2018 by American Association of Geographers, reprinted by permission of Informa UK Limited, trading as Taylor & Francis Group, www.tandfonline.com, on behalf of the American Association of Geographers. An earlier version of chapter 4 appeared as "Governing a Liminal Land Deal: The Biopolitics and Necropolitics of Gender," *Antipode* 52, no. 3 (2020): 722–41, copyright © 2020 by John Wiley and Sons. Portions of chapter 6 appeared as "The Curious Case of Three Male Elders: Land Grabbing, Lawfare, and Intersectional Politics of Exclusion in Tanzania," *African Studies Review* 64, no. 3 (2021): 605–27, reproduced with permission by Cambridge University Press. I wish to thank the reviewers and editors of those journals at the time of publication (James McCarthy, Kiran Asher, Benjamin Lawrence) for their critical feedback and guidance.

I am sincerely grateful to the editors of the Cornell Series on Land, Wendy Wolford, Nancy Lee Peluso, and Michael Goldman, for their unwavering support for my book. I am particularly indebted to Nancy for her honest and incisive feedback and for always being available for advice in-person and over the phone, text message, and email. I also thank Jim Lance, Clare Jones, Mary Kate Murphy, and other editorial staff at Cornell University Press for shepherding this book into production. The two anonymous readers of the manuscript provided the most detailed, perceptive, and compelling reviews I have ever received on my writing. They saw right through the portions of the manuscript where I ran out of steam and pushed me to be clearer and more coherent in my use of theory, while making the book accessible to diverse audiences. Kathleen Kearns's meticulous developmental editing, which involved just the right balance of criticism and affirmation, helped advance the first full draft of this book; I owe Andrew Ofstehage for introducing me to her. Many thanks to Christine Riggio for polishing and significantly improving the legibility of my maps and to Glenn Novak for thorough copyediting. Although our paths diverged, I would also like to acknowledge Amy Trauger and Jennifer Fluri, the editors of the Gender, Feminism, and Geography series at West Virginia University Press for their support, as well as the anonymous readers they solicited to review my book proposal.

My family has been my biggest fans, even if they did not always understand the particularities of academia and my research. My parents, to whom this book is dedicated, instilled in me the value of education and hard work since an early age and nurtured independence, creativity, and compassion for others. My sister, brother-in-law, and nephew kept me honest about my priorities, especially during the last two years of intense writing. My maternal grandparents, who cared for me in my early childhood years in Seoul, were proud to see me embark on an academic journey. They witnessed and lived through colonialism, war, dictatorship, developmentalism, and neoliberal globalization all within their lifetime. I wish I had the chance to learn more about their life and for them to see this work come to fruition. The Doll family has shown me nothing but love since the day I met them. They never failed to ask how the book was going, and I am delighted to finally be able to share it with them. Two important family members and contributors to this book have not read it. Wall-E accompanied me on this journey since the beginning of my PhD, and both he and our newest addition, Leo, have sat on my lap, purring their little motors, as I wrote paragraph after paragraph, day after day. Most of all, I am profoundly grateful to my life partner and husband Ross. He has read countless versions of my writing with the utmost care and thoughtfulness. His brilliant intellect and warm companionship have sustained me throughout the years of writing this book. He cheered me on every day, tempered my perfectionist tendencies, and always reminded me to smell the roses. This is as much his accomplishment as it is mine.

Abbreviations

AfDB	African Development Bank
ASDP	Agricultural Sector Development Program
BRN	Big Results Now
CCM	Chama Cha Mapinduzi (Party of the Revolution)
ESIA	environmental and social impact assessment
EU	European Union
GTZ	Deutsche Gesellschaft für Technische Zusammenarbeit (German Agency for Technical Cooperation)
ICSID	International Centre for Settlement of Investment Disputes
IFAD	International Fund for Agricultural Development
IFI	international financial institutions
LBS	Leipziger Baumwollspinnerei (Leipzig Cotton Mill)
MP	Member of Parliament
NGO	nongovernmental organization
PAP	Project Affected Person
PMO	Prime Minister's Office
PPC	People and Property Count
PPP	public-private partnership
RAZABA	Ranchi ya Zanzibar Bagamoyo (Ranch of Zanzibar in Bagamoyo)
SAGCOT	Southern Agriculture Corridor of Tanzania
SBSA	Standard Bank of South Africa
SEK	Swedish krona
Sekab	Svensk Etanol Kemi Aktiebolag (Swedish Ethanol Chemistry Limited Liability Company)
Sekab BT	Svensk Etanol Kemi Aktiebolag BioEnergy Tanzania
Sida	Swedish International Development Cooperation Agency
TANAPA	Tanzania National Parks Authority
TANESCO	Tanzania Electric Supply Company
TANU	Tanganyika African National Union
TIC	Tanzania Investment Center
TYL	Tanganyika African National Union Youth League

TZS	Tanzanian shilling
URT	United Republic of Tanzania
USAID	United States Agency for International Development
USD	United States dollar
WWF	World Wildlife Fund

Note on Currency

This book makes references to figures in US dollars (USD) and Tanzanian shillings (TZS). Unless otherwise noted, all currency conversions are based on the historical yearly average rate between 2012 and 2022, where USD 1 was exchanged for around TZS 2,000.

**SWEET DEAL,
BITTER LANDSCAPE**

INTRODUCTION

"Bagamoyo was upbeat," opened an article in the Dar es Salaam–based *Citizen* on March 17, 2014.[1] The town was set to embark on the early stages of what the article called a "model sugar production programme" and Tanzania's "pilot" transnational land deal, which had been delayed since its inception in 2005.[2] It was not every day that Bagamoyo made national headlines. Once a lucrative coastal trading entrepôt for the Omani Sultanate of Zanzibar and the capital of German East Africa during the nineteenth century, it had since then declined in commercial significance and had gained the popular epithet of a sleepy town. The appearance of major dignitaries for the launch event, including cabinet ministers, foreign diplomats, and development agency representatives, portended an economic renaissance for the town and district of Bagamoyo.

For the land deal, known as the EcoEnergy Sugar Project (formerly Sekab BioEnergy Project), the state had transferred approximately 20,400 hectares of land (50,400 acres, or about 79 square miles) under a ninety-nine-year lease to a Swedish company that promised to mobilize one trillion Tanzanian shillings (USD 500 million) for commercial sugarcane production. In exchange for offering land "free of encumbrances," the Tanzanian government would receive up to 25 percent equity shares in the project company; this type of financial sharecropping had no precedent in the country or the East African region as a whole.[3] As the nation's first new sugarcane plantation to be established in over forty years, the project was envisaged to achieve many ambitious goals: to produce, on an annual basis, 150,000 tons of sugar to resolve the national sugar deficit, 12,000 cubic meters of ethanol to mitigate global climate change, and 90,000 megawatt

1

hours of electricity to support Tanzania's energy security, while creating twenty thousand new jobs, including the training of up to two thousand local farmers as commercial outgrowers. Promoted as a public-private partnership, a new model for international development that has grown in popularity, the project also received promises of funding support from the African Development Bank (AfDB), the United Nation's International Fund for Agricultural Development (IFAD), and the Swedish International Development Cooperation Agency (Sida).

In a ceremonial gesture to signify new beginnings, the high-profile guests at the event each planted a cane cutting in a big pot of soil prepared on the stage in front of a large crowd. Noting that the project was potentially the largest of its kind in East Africa, EcoEnergy's executive chairman expressed his commitment to bringing it to fruition: "When you unleash such a big project . . . you cannot stop in the middle of it." After ten years of negotiations and delay, the "early works" on the project would finally begin (figures 0.1 and 0.2). These would include land clearance and infrastructure construction; compensation and resettlement of local populations; establishment of site offices and project signages; and increased security measures like the deployment of paramilitary forces (*mgambo*) with the support of the district government.

In his remarks, the Swedish ambassador to Tanzania emphasized how the project would "break new grounds" in the fifty years of "friendship" between the two nations—a relationship originally built on their ideological affinity to

FIGURE 0.1. A sign shows a map of the project area and states in Swahili that the project is a partnership between the Tanzanian government and EcoEnergy. Photo by the author, August 2014.

FIGURE 0.2. Concrete posts on the western border of the EcoEnergy project area were installed in 2014 as part of the project's early works. Photo by the author, March 2016.

socialism in the wake of decolonization.[4] Reaffirming the ambassador's words and preempting potential critics among the crowd, the minister for agriculture stated in his remarks: "There is an issue that has been debated everywhere globally, and that is land grabbing. . . . There is no land grabbing in Tanzania. . . . Investors do not come here to replace smallholder farmers." Instead, the EcoEnergy project, the minister said, would mark an important milestone in President Jakaya Kikwete's ongoing effort to bring agricultural modernization and rural development into a "win-win" partnership with foreign investors.[5]

A handful of villagers were present at the event, most of them from outside the concession area and known supporters of the project. They were farmers from neighboring villages who had been participating in the company's nascent outgrower and community development program. A leader and several members of a farmers' group, mostly men, stood at the podium, all wearing a black baseball cap embossed with the blue EcoEnergy company logo; local residents would later consider anyone with that hat as one of EcoEnergy's *chachu* (provocateurs; literally, yeast). Fondly recalling a company-sponsored field trip to Kilombero District to observe an already existing sugar plantation and outgrower scheme, the male farmer leader endorsed the project and urged smallholders in the district to organize themselves to realize the value of land.[6] "Our land," he said, might in fact be "gifted with oil," referring to the economic potential of sugarcane and its many derivatives, including ethanol and the EcoEnergy project at large.

Notably absent from this event were hundreds of families who had been scheduled for removal from the concession area. According to the project's 2012 executive summary of the Resettlement Action Plan, the land deal would affect a total of 1,374 people. Despite the semblance of precision, this number was contestable.[7] The scope of the 2012 plan was limited to only those who would be displaced in the first phase of land clearance, in the northern part of the project site along the Wami River, where most irrigation systems would be concentrated; plans for the subsequent three phrases with updated displacement figures were never published. Local residents disputed the accuracy of the reported head count, arguing that some people were not included, or inconsistently included, in the first round of the census that government authorities and project consultants had conducted in 2011. Those omitted included families who lived far away from the main roads and were thus difficult to reach, individuals who refused to be counted, as well as wives and children, including those born after the census, who were subsumed under the identities of their husbands and fathers. Some semi-nomadic pastoralists were counted in the census but were presumed to own no land, permanent dwellings, or other assets considered worthy of compensation.[8] A district land officer I interviewed who had been involved in the second round of census partially completed in 2014 gave me a higher estimate, around three thousand people.[9] Regardless of their numbers, status, or eligibility for compensation, the project promised to allow all current occupants to "harvest and leave" before land clearing commenced.

Six months after the commencement of the early works, the company had yet to clear a single hectare of land. The project delay persisted for many more months, and by April 2015 the journalist who had covered the "upbeat" launch event a year prior declared that the land deal had "vanished into thin air."[10] Then, unexpectedly, in May 2016, the prime minister announced to the Parliament that the deal was being canceled to protect the well-being not of the local residents but of the wildlife in an adjacent coastal national park.[11] Notwithstanding the ambiguities surrounding the prime minister's announcement, a freelance correspondent for the *Thomson Reuters Foundation News* claimed that villagers in Bagamoyo had won a "rare victory" by being "spared eviction" from a transnational land deal.[12] To other observers, including government officials and donor agency representatives I interviewed, the land deal had simply fallen through, with no immediate or long-term consequences.[13]

This apparent collapse or failure of the land deal, however, belied the confusions, tensions, and struggles that continued to shape everyday life on the land. A tumultuous decade in the making, the land deal had cast a long shadow over people's lives—a shadow that would not fade easily with the news of its presumed disappearance. Hardly dead, the land deal was still very much alive in people's minds and lives. Hardly spared eviction, people were scrambling to hold on to their land and livelihoods, while trying to make sense of what it meant to live in a spacetime of liminality, waiting for development and dispossession yet to come.

This book is about the indeterminate trajectory of the EcoEnergy land deal and how it shaped social life on the land in myriad gendered ways for diverse rural people in coastal Tanzania. The case of the EcoEnergy Sugar Project—how it came about and became stalled, and how it was governed and experienced on the ground—raises critical questions about the social dynamics of late agrarian capitalism. The EcoEnergy case shows how corporate investors, states, and donors are simultaneously implicated in the plunder and management of rural resources and landscapes, but cannot always act as they please or under the circumstances of their own choosing.[14] While this book offers several explanations of why the project unfolded the way it did, I argue that it fundamentally could not take root because it became deeply enmeshed in Bagamoyo's bitter landscape. As I will show, this landscape was shaped not only by gendered cultural, material, and ecological processes but also by people's persistent struggles to remain on the land despite repeated cycles of enclosure from colonial times to the present. Corporate and state actors, entangled as they were in this historically contested landscape, had to work to control both people and resources to facilitate rural dispossession. Their mechanisms of control, which involved a combination of consent and coercion, were inconclusive in their results, but they nonetheless produced real social effects.

I focus specifically on the ways that gender, race, class, and other intersecting inequalities shaped and were shaped by the project's governance during the years of delay. I also investigate how these social relations of power informed the multiple and sometimes contradictory ways different groups of women and men resisted the increasing pressures on their livelihoods and negotiated the seemingly endless liminality—a sense of being in-between, about to happen, profound and quotidian uncertainty—that permeated daily life. In tracing the everyday politics of survival and social reproduction on the margins of new and incomplete enclosures in Tanzania, the book sheds light on the importance of history, place, and power in the interconnected trajectories of rural development, postcolonial nation building, and neoliberal globalization in Africa and beyond.

When I first arrived in Bagamoyo in 2013, I intended to study displacement as it happened. I wanted to follow the families the EcoEnergy land deal was displacing *ex situ* and understand how this process reshaped their relationships to the land and resources, as well as the gendered dynamics of production and social reproduction in the resettlement areas. When I returned in subsequent years, it became evident that what had seemed at first glance imminent displacement was becoming more and more elusive, but no less real. People were relieved to still be on the land, but they weren't sure how much longer they could remain. So much was unknown. One thing they knew for certain was that their lives had become inextricably bound up with "the project" (*mradi*). They were in the process of being displaced *in situ*, an experience of slow violence in which people remain in place but with diminishing access to resources, an eroding sense of belonging, and increasing pressures on livelihoods.[15] As many people described and illustrated in their photo-narratives, living under the project's shadow was like "living with one foot in, one foot out" or "living in parentheses." It was as if they were "living like refugees," or being "put in a cage" or "squeezed between brackets" (figures 0.3–0.5).

> When the project people came and asked if this land was mine, I said of course, look at this tree! When my neighbors see this tree, they know that they are on my land. But the project is giving us a lot of worries. Investment isn't bad, as it will benefit the nation. But the way they are doing things, it makes me feel terrible. For so many years, we have been told not to grow things with roots, not to build houses, not to expand our farms. This is no way to live. The land is ours, but we are living here like refugees.[16]

FIGURE 0.3. A tree carving. Photo and narrative by Athumani, male farmer, January 2016.

By the time I left Bagamoyo at the end of 2016, this embodiment of in-betweenness had become a new normal. It had become part of the fabric of social life. Liminality was not a fleeting impression. It was, at once, a sustained and suspended experience of change, although the specific ways in which different individuals and groups registered and acted upon it varied. The kinds of change people experienced on the ground, such as land use restrictions, threats, and violence the *mgambo* inflicted, were not spectacular or revolutionary per se. They were mundane, inconsistent, and invisible from public view, but they nonetheless signaled change with deep material, social, and affective consequences.

> I worry about getting caught by the *mgambo* every time I go collect fuelwood. I pray to God often. People are finding it hard to move forward with their lives. It feels like we are being squeezed, like we've been put in between brackets. It's like we've been informed of an impending death of a relative. Life goes on, but it's difficult not knowing when or where I will be moved to. . . . Men can always go off somewhere and do casual work. It's harder for women, older people, and especially widows like me. And much harder for elderly women who are disabled, like my mother.[17]

FIGURE 0.4. A pile of fuelwood. Photo and narrative by Neema, a female farmer, August 2016.

We are frustrated. We are tired. We are losing our minds. It's like we've been put in a cage. They put *X* marks on our houses and trees. We are only allowed to grow maize, nothing else. If it weren't for this project, we would have mangos and jackfruits by now. Land is our only wealth; farming is our livelihood. We are asking ourselves: how did we become so poor?[18]

I use liminality as a guiding heuristic to organize my analysis and to refer to the socially constructed condition and experience of in-betweenness, specifically as it relates to the unpredictable spacetime between land acquisition and capital accumulation. As a lived and felt phenomenon, liminality indexes what cultural critic Raymond Williams called "structures of feeling," or the "complexities, experienced tensions, shifts, uncertainties, the intricate forms of unevenness and confusion" that are inherent to the process of change but that social analysis often misses or brackets as noise.[19] Liminality, as I go on to show, is not an accidental or static condition where everything is temporarily "on pause." Rather, it is a dynamic and contingent spatiotemporal relationship that state, corporate, and various nonstate actors with divergent and incommensurable interests, values,

FIGURE 0.5. A tree marked for removal by two red *X*'s. Photo and narrative by Nuru, a female farmer, September 2016.

and subject positions coproduce, maintain, contest, and at times exploit. While the condition of in-betweenness may be inherent to life itself, the liminality associated with the EcoEnergy land deal must be located in the geohistorical conjuncture of neoliberal globalization and postcolonial postsocialism in Tanzania. This introduction situates the land deal in these broader contexts and outlines the book's contributions for both critical theory and politics.

Liminality of Development and Agrarian Change

The case presented in this book and a number of examples like it across the globe unsettle the teleological view of history assumed in dominant narratives of development and modernization. The perception that all human societies evolve in a linear progression from "primitive" to "modern," from "agricultural" to "industrial," has had long roots in Western thought and philosophy that naturalized slavery, imperialism, and colonialism and that gave birth to the positivist social sciences in the early nineteenth century.[20] The idea also influenced the classic late nineteenth- and early twentieth-century Marxist thinking around the "agrarian question," which asked how capital takes control over peasant agriculture and "backward" rural areas—although the political assumption here was that socialism would supplant capitalism as the end of history.[21] In the context of decolonization and the Cold War, the apparatus of international development, or Development with a "big D" as designated by Gillian Hart, came to signify the global application of modernization theory by a whole gamut of governmental, intergovernmental, and nongovernmental organizations to improve the societies, economies, and ecologies of formerly colonized or "underdeveloped" nations.[22]

The EcoEnergy land deal was born out of these abiding modernist beliefs about the inevitability of capitalist agrarian "transition" and the paternalism ingrained in the project of "Development." The foreign investor and the Tanzanian state envisioned the land deal—and major donor agencies supported it—as a necessary step toward improving Tanzania's competitiveness in the global economy, supporting its upward path to becoming a middle-income country by 2050.[23] As the Swedish EcoEnergy corporate executive said at the launch event in 2014, a project of this magnitude and significance could not possibly be stopped. A British engineer and contractor for EcoEnergy's outgrower and community development program expressed a similar sentiment. When I asked about his vision for the outgrower scheme in an interview, he stressed the urgency of helping Bagamoyo's "backward" smallholder farmers transition from their current "survival mode" to "grow mode." By joining the project, he argued, these farmers will finally be "entering the modern world of commercial agriculture" and be "winning and uplifting themselves."[24] The EcoEnergy land deal exemplifies what critical scholars have been calling the "new enclosures" of the twenty-first century. In Tanzania and elsewhere, such deals are giving new life and relevance to the "old" agrarian question. The question is made more salient when we recognize its convergence with the coloniality of "Development."[25]

If modernization thinkers were correct, all the contracted land in Bagamoyo would have been fully privatized, and rural Africans would have been dispossessed of their means of production with nothing but their labor to sell to the sugar plantation, factory, and/or other low-wage employers. Yet, more than ten years after the project's inception, enclosure remained simply incomplete; it was partially realized on paper but yet to be realized on the ground. There was no bloody "primitive accumulation" of the kind that Marx had described, despite other forms of in situ displacement and violence already mentioned (and to be discussed further in chapter 4). Instead of unfolding in a preordained linear progression, the land deal encountered multiple twists and turns, and its fate and long-term repercussions still remain in question. To better understand and explain this "myth of modernization," to borrow from James Ferguson, it seems necessary, then, to turn the original agrarian question on its head: How and under what conditions is capital *unable* to seize control over agriculture and the countryside, and with what consequences?[26] What concepts and methods are appropriate for explaining indeterminate agrarian transformations and for disrupting capital's telos?

These questions are not just pertinent to the EcoEnergy case. Increasing evidence suggests that many of the large-scale land deals signed during the first decade of the twenty-first century have either been stalled, abandoned, canceled, or at least significantly downsized.[27] According to a study published in 2016, of approximately 22 million hectares (about 85,000 square miles) investors contracted for agriculture in Africa between 2000 and 2014, only about 3 percent (0.7 million hectares) was under cultivation.[28] Figures from the Land Matrix, an online global crowdsourced database, further illustrate this trend. As of April 2022, there were estimated to be approximately 150 "intended" land deals on 13 million hectares globally, or deals where contracts have been signed but whose implementations have been delayed for many years. This is in addition to roughly 240 "failed" land deals on 27 million hectares, where contracts have been terminated or land titles revoked.[29] Though the definitions and boundaries between these categories are fuzzy, available data make clear that the vast majority of these unfinished land deals are located in sub-Saharan Africa—a region that had received the greatest investor interest in the early days of the global land rush, igniting concerns about a "new scramble for Africa."[30]

Examples of these partial enclosures abound in Tanzania. In addition to the EcoEnergy land deal, Bagamoyo District has been home to another suspended transnational land deal, for the development of the largest deepwater port in Africa as part of China's Maritime Silk Road Project and the broader Belt and Road Initiative.[31] Even major land deals that began some operations on the ground, such as the British jatropha and rice plantations in Kisarawe and Kilombero Districts (Sun Biofuels and Agrica) and the Emirati hunting concession in Ngorongoro District (Ortello Business Corporation), all experienced a halt at one point or another when faced with compounding challenges, including land use conflicts, bankruptcy, social mobilizations, court injunctions, and allegations of corruption—all elements that also gripped the EcoEnergy project.[32] From Madagascar to Mozambique to Ethiopia to Senegal, similar examples abound across the continent.[33] Observing this trend, Timothy Wise, a senior adviser at the Institute for Agriculture and Trade Policy, wrote in 2014 that "failed land grabs litter the African landscape."[34] Ten years after sparking the "global land grab" debate in 2008, GRAIN, a Barcelona-based nongovernmental organization, also began turning its attention to "failed farmland deals."[35]

In this book, rather than classifying as failures the land deals that did not go the way project planners had intended, I find it more generative to think with the concept of liminality. Whereas failure implies a done deal characterized by absence, obsolescence, closure, or inability to affect, liminality, with roots in the Latin word *limen*, meaning threshold, leaves room for contingency, ambiguity, open-endedness, and the possibility of politics. Choosing to think with liminality is more than a matter of semantics. Characterizing land deals that did not materialize as planned as failures can afford a sense of urgency and polemical punch to global campaigns against land grabbing. And many critical scholars, to date, have usefully analyzed the routine so-called failures of development projects and their instrumental effects, such as the depoliticization of poverty and the strengthening of bureaucratic power, central to biopolitical governmentality and domination.[36] Yet defining the EcoEnergy land deal as a failure would do injustice to the fact that the people on the ground did not think of it or experience it that way, even as the state and the media were quick to describe it as such. Rather than starting my inquiry "outside experience and in the discourse," to borrow from feminist sociologist Dorothy Smith, I situate the ordinary experiences of liminality—as diverse as they are—at the heart of my inquiry.[37]

As an analytic sensibility, liminality allows for ethnographic investigations that attend to what goes on in the margins, that disrupt the binary narratives of success and failure endemic to development, and that reveal the diverse, power-laden expressions and experiences of uncertainty. In his early use of the concept, anthropologist Victor Turner had defined liminality as "the domain of the 'interesting,' or of 'uncommon sense.'"[38] Rather than being some meaningless lull where nothing happens, it is precisely where things happen, but in unexpected ways.[39] Feminist and queer theorist Gloria Anzaldúa reminds us, too, that change or the possibility of change often emerges from these in-between spacetimes that are unstable and ambiguous, lacking in clear boundaries.[40] Other postcolonial theorists have discussed liminality as constituting a disjunctive realm, characterized not only by a collapse of certainty but also by an active renegotiation of identity and agency for subjects of oppression and dispossession.[41] Ultimately, I suggest that studying incomplete land deals with a heuristic of liminality offers an opportunity and a language with which to reframe and critique development, not as a smooth inevitable unfolding but one that is immanently contingent and indeterminable, shaped by many interstitial and discontinuous processes and unequal power relations.[42]

As I sketched above and will return throughout the book, the EcoEnergy land deal produced its effects not in spite of but because of its liminality, especially in its formative years between 2005 and 2015. Reports of its ostensible failure concealed the ways in which it shaped social life on the land in immeasurable ways during this period, as well as after the abrupt end of the contractual relationship between the government and the investor in 2016, which I delve into in the conclusion. This pattern is seen elsewhere, too. Available case studies suggest that when land deals are terminated, land is seldom returned to its former owners; often it remains with the state or is reallocated to other investors or for other purposes, creating further confusions and conflicts over land tenure.[43] Special-purpose project companies that are established as a result of land deals may be sold to other overseas owners, or, as I will discuss in the conclusion, foreign investors like EcoEnergy may resort to international arbitration, another long and unpredictable process that, by design, excludes the voices of rural communities. In sum, land deals last a long time. They take on lives of their own, often improvising and adopting makeshift adjustments and coming up against various frictions as they encounter local and regional particularities and historical exigencies.[44]

Landscapes of Power: Gender, Intersectionality, and the New Enclosures

EcoEnergy's decision to site the project in Bagamoyo was initially based on an assessment of satellite images of mainland Tanzania's two-hundred-kilometer coastal strip.[45] Taken from thousands of feet above the terrain, these visualizations helped create an illusion of an empty landscape: a *terra nullius*. What the images generalized as vegetation or tree cover obscured the myriad indigenous tree species rural women and men gathered for food, fuel, fiber, medicine, building material, and cultural rituals.[46] What the images displayed as a river flattened the rich histories and ecologies of floodplain agriculture and fishing activities that defined the livelihoods of coastal peoples.[47] The identification of a railway as a natural boundary masked the sweat and racialized labor of the local and migrant men who built it during the late colonial and early independence period. When I asked a senior Ministry of Lands official about existing patterns of land use in the concession area, he quickly defended the objectivity of aerial photographs: "There is no land dispute. There is no contradiction. I should show you the satellite images. The land has been invaded; people who are living there now are doing so illegally."[48]

In the transaction between the Tanzanian government and EcoEnergy, land was a portion of the earth's surface that was valuable insofar as it could be alienated and extracted for rent, interest, dividend, and profit. An investment promotion presentation the government produced in 2012 highlights cheap land and labor as chief among the nation's unique selling points. Lease costs (in USD) would be "less than $1/ha/year, after initial compensation to any land users" and labor costs, "about $180/month including taxes and housing allowance."[49] Investors in the global land rush considered Africa to be a "good place to find scale," because in countries like Tanzania, not only was land comparatively cheap but also the state was the sole landowner.[50] The Tanzanian land laws, as I elaborate in chapters 1 and 2, vest ultimate land title in the president on behalf of all citizens, and this title comes with the prerogative to exercise compulsory acquisition of village lands and transfer them to investors for projects of "public purpose."[51]

But even when investors have obtained de jure rights to land, this does not directly translate into de facto access to it.[52] Access, according to Jesse Ribot and Nancy Peluso, refers to one's ability to derive benefits from resources, as mediated by a whole range of social relationships or "bundles and webs of powers."[53] The discrepancies between rights and access and between investors' aspiration and reality exist precisely because enclosures entail more than the technicalities of commodifying nature. They involve major transformations in world-making practices and lives lived on the land. As the above promotional presentation implied, land would be "cheap" only insofar as an indeterminate number of existing resource users have been paid compensation to voluntarily abandon their land. However, land is not a self-contained thing from which one can easily remove oneself. Its social and physical boundaries are porous and intertwined; land is always more than the sum of its parts. For those who work it, it is not just a means of production but a home and a vital source of social reproduction.[54] In rural Bagamoyo, the lived histories, memories, ecologies, and cultures overflowed what was precisely measured out to be 20,373.56 hectares. The task of controlling these hidden overflows, I argue, would continue to haunt the project's quest to transform Bagamoyo into a field of "white gold."

Considering land's polyvalent qualities, the book conceptualizes contemporary land deals as enclosures of rural landscapes. Drawing on the insights of political ecology and more-than-human feminist perspectives, I define landscapes as the material and symbolic result of people's ongoing relationships with nonhuman natures (land, water, river, floodplains, crops, trees, forests, weeds, wildlife, and the like) while being intimately tied to extralocal political economic processes. Landscapes, thus, are outcomes of everyday presence, practice, and politics. They are produced through people's daily and generational struggles over resources and meanings, including the definitions of identity, belonging, and citizenship.[55]

Embedded in my analysis of enclosures is a feminist critique of capitalism and the production of nature, which have historically gone hand in glove with the domination of gendered and racialized bodies and the devaluation of the feminine.[56] As I will demonstrate, governing the EcoEnergy land deal required control over both resources and people, an undertaking that manifested itself in important gendered ways. To explain the social construction of gender, Judith Butler has usefully defined it not as a thing or a set of universal attributes, but as a set of acts that is "performatively produced and compelled by the regulatory practices of gender coherence."[57] Gender, in other words, is both embedded within and productive of power structures that shape the contours of hegemonic femininity and masculinity.[58] While I use gender as an anchor to guide my analysis throughout the book, it must always be understood as an articulated category.[59] The meanings and experiences of gender are inextricably bound up with those of race/ethnicity, class, age, religion, sexuality, ability/disability, and gradations of social status, such as those shaped by one's marital status, residential status, and educational attainment. I list examples of differences here not to debate whether one is more or the most important (in an "Oppression Olympics," as Ange-Marie Hancock has called it), but to highlight the inherent complexity in studying gendered power relations.[60] The ways in which gender intersects with other axes of power must also not be assumed, but rather be historicized and empirically grounded in particular historical and geographical contexts.

A historical feminist political ecology perspective would indeed lay bare how every cycle of enclosure in Bagamoyo and coastal Tanzania has depended on and reproduced what decolonial feminist philosopher Mária Lugones, drawing on Aníbal Quijano, has usefully termed "the coloniality of gender."[61] During the nineteenth-century Indian Ocean slave trade, African women and girls were captured and turned into staple commodities, with their bodies valued for their perceived utility in domestic servitude, concubinage, and agricultural labor in

Arab plantations.[62] They were often seized when they were young, sometimes pawned by their own male elder kin, and frequently sold from one master to another until they became wholly alienated from their lands. On the eve of European colonialism, French Catholic missionaries in Bagamoyo seized land from indigenous territories through what historian Walter Brown described as "a unilateral policy of aggrandizement," and these missionaries later became one of the largest property owners in the region.[63] In proselytizing Christianity to the predominantly Muslim populations of the Swahili coast, missionary stations and schools normalized the separation of women and men into dichotomous spheres (public/private, political/domestic), building on what were presumed to be traditional gender norms in which women and girls were deemed slaves, commodities, and properties of men.[64] Colonial rule reinforced these gender ideologies by imposing the Victorian ideal of a patriarchal nuclear family through male migrant labor and taxation policies, all of which subordinated wives under the political and economic authority of their husbands.[65] During this period, European colonial authorities "domesticated" African women as the mainstay of the countryside and largely prohibited them from out-migrating to the cities; if they failed to meet the required production targets according to the minimum acreage bylaws, they were subject to penal sanctions, including imprisonment.[66] When allowed paid employment, women were relegated by the colonial administration to domestic service based on the sexist notion that housework and care work were "unproductive."[67] In the late colonial period, the government worked to codify various living cultural practices that governed matters of the family, marriage, property, and inheritance into a single customary "law." This process was informed by studies a male European anthropologist conducted with the assistance of male colonial officers and male elders of select patrilineal ethnic groups, which resulted in a highly patriarchal version of customary law.[68] This "patrilineal offensive"—which was not uncommon across the colonial world—had the effect of delegitimizing the diversity of living customs of over 120 ethnic groups in Tanzania, including matrilineal ones in coastal regions.[69] The Customary Law (Declaration) Order was enacted in 1963, shortly after Tanzania gained independence, and despite persistent calls for legal reforms, they remain in effect today without significant amendments.

The point of this brief review is not to offer a historical "backdrop" but to underscore how patriarchy and the construction and regulation of gendered identity have been inseparable from the very dynamics of enclosure, and the consequent governance of property, the nation, and the family. This means that any study of contemporary enclosures, whatever trajectories they might take, would be incomplete without feminist analysis of governmentality, intersectionality, and subject formation. Single-axis (e.g., class-centric) analysis, therefore, is insufficient for understanding both the structural dynamics and the everyday lived experiences of late agrarian capitalism.

The Rural, the Nation, and the Politics of Resistance

The third contribution of the book is that it highlights the centrality of agency, or the varying degrees and expressions of agency of rural actors, in the story of new enclosures. Those who appear in the pages of this book are not passive homogeneous actors caught powerlessly in an inevitable transition toward depeasantization and proletarianization. They are active heterogeneous agents who engage in a whole host of strategies to make do, get by, and contest their shifting conditions of life on the land.

The category of the rural farmer/villager, however, has occupied an ambiguous position in Tanzania's history. On the one hand, the political imaginary and narrative of peasants who work the land have been pivotal to Tanzania's nation-building project since independence, resurfacing again and again at every election cycle. As Julius Nyerere, the Father of the Nation (*Baba la Taifa*) wrote in 1968, "The land is the only basis for Tanzania's development; we have no other.... Tanzanian socialism must be firmly based on the land and its workers."[70] Tanzania's program for African socialism, known as *ujamaa* or familyhood, emerged between the late 1960s and early 1970s, drawing on several key themes of Maoism, namely self-reliance, mass mobilization, and pro-peasant politics.[71] Most notably, ujamaa involved a profound reorganization of the countryside and the resettlement (first voluntary, but later forced) of millions of peasants into socialist villages so they could work collectively to "build the nation" (*kujenga taifa*).[72] Propagandistic materials from the period depicted the hoe as a symbol of the nationalist project;[73] and the hoe still remains, alongside the hammer, on the flag of the nation's ruling party, Chama Cha Mapinduzi (CCM), or Party of the Revolution—the longest-reigning ruling party in Africa—as a cultural remnant of the imagined socialist community rural peasants and urban workers were together supposed to construct.[74] Whereas Marx and early Marxist thinkers had assumed that European peasants were apolitical, homologous masses who were destined to disappear with the march of modernity (Marx's proverbial "sack of potatoes"), African peasant farmers were expected to contribute to the socialist revolution in and from the villages through collective agriculture.[75]

But the implementation and outcomes of this idealized version of African agrarian socialism were highly context-specific and often contradictory.[76] In Bagamoyo and other coastal districts, people frequently refused to work in communal farms and eventually returned to their pre-ujamaa land use practices after compulsory villagization came to an end in the late 1970s, as I discuss in chapter 2. At a broader level, the state's understanding of familyhood reinforced the inherited colonial ideals of the nuclear family and domesticity, such that the division of socialist developmental labor became defined in rigidly gendered terms: women were expected to perform the majority of agricultural and reproductive labor to ensure household food security and well-being, while men, especially young militant men, defended national security by performing military service and policing duties, including the enforcement of compulsory villagization—the legacies of which I discuss in chapter 4.[77] Beyond naturalizing gender difference, party-state leaders also sanctioned the exclusion of racial-ethnic minorities on the grounds of cultivating a common African national identity.[78]

Since the abandonment of ujamaa and the imposition of neoliberal economic restructuring in the late 1980s, however, the once-patriotic figure of the hoe-wielding peasant has gradually lost its salience. While policy makers still consider agriculture a key pillar of the national economy, their discourse has increasingly portrayed small-scale food producers as obstacles to progress. In calling attention to the role of "spatial identities" that transcend gender, racial, and ethnic identities, anthropologist Kristin Phillips has argued that the socioeconomic marginalization of "the rural" has been foundational to statecraft and nationhood in postcolonial- postsocialist Tanzania.[79] As she writes, the struggles for citizenship and national belonging in Tanzania often "take place within a social field that allots rights, responsibilities, and resources partly in accordance with a distinction between city and the village."[80] Building on this work, I suggest that the question of the rural is indispensable to understanding the contradictions of capitalist development and nation building in the statist-neoliberal moment, where political elites increasingly seek to promote economic growth through large-scale agricultural and extractive investments, all the while trying to secure their legitimacy from poor rural citizens, who not only constitute the vast majority of the nation's voters but also depend on the very land and resources the state and investors need for their livelihoods.[81]

The EcoEnergy case epitomizes this ongoing dialectical tension between capital accumulation and political legitimization and the way the rural figures centrally as a terrain of struggle. Recall the agricultural minister's remarks at the project launch event in 2014. Although he reassured the crowd that investors weren't coming to Tanzania to make smallholder farmers redundant, he nevertheless implied that the key to the nation's so-called green revolution lay not in subsistence but commercialization and incorporation of small-scale producers into value-chain agriculture and debt relations through contract farming.[82] In Bagamoyo, the presumed inefficiency and fungibility of the rural poor, multiplied by the state's ultimate landownership and historical amnesia of past enclosures and their ongoing legacies, have given the state license to represent all residents within the concession area as "intruders/squatters" (*wavamizi/waingiliaji*), but not without contention. As one villager told me, "No choice was given to us. The only option we were given was to leave. But if we were truly intruders, why are politicians coming to us with their campaigns?"[83] Another complained, "We have been stripped of our humanity. The MP of Bagamoyo said, 'You are intruders. People will piss on your graves. You do not belong here. So leave.' But then he came back later when he needed our votes. He said, 'You are my people. I will help you get your land back. Please vote for me.'"[84]

How useful is it—politically, theoretically, and ethically—to position rural subjects as being inconsequential, nonbelonging, and bereft of choice, except during the election season? Writing about the ambiguous position of plantation workers cum peasants who produce food on Colombian plantation margins, Michael Taussig described them as "liminal beings . . . neither what they are, or what they will become."[85] While the rural women and men living within EcoEnergy's would-be plantation similarly expressed a feeling of living in limbo—constantly made to feel like "refugees" who did not belong either to their land or the nation—they were not, in any sense, lacking in subjectivities, aspirations, and moral expectations, as Taussig's description tends to conjure. If we understand the liminality of the land deal and the coupled liminality of agrarian life in Bagamoyo as outcomes of power, we can safely reason—drawing on Gramsci, Polanyi, and Foucault—that they would be met with reactions, countermovements, or resistances of various kinds.[86]

In this study, I draw on the rich tradition in critical agrarian studies to examine a wide array of practices—both hidden and overt, mundane and dramatic—with which different groups of women and men resisted living under the shadow of the EcoEnergy land deal.[87] In studying subaltern agency with ethnography's commitment to "thick descriptions," however, the book shows how the category of resistance can sometimes obscure the gendered and intersectional dynamics of agrarian politics.[88] There is no such thing as "local people," "households" or "communities" without internal politics and differentiations.[89] Although strategic essentialism has at times enabled the mobilization of otherwise complex identities in political struggles of disenfranchised groups, the convenience of simplifying the subaltern comes at the expense of understanding how people's divergent social positions and positionings may in fact hinder rather than facilitate opportunities for coalition building and collective action.[90]

As I discuss in chapter 6, eliding difference can also obscure events that may pass as resistance, or be justified by certain elite actors as resistance, but which ordinary people might perceive as distortions of it. Without attending to these nuances, we risk tethering the meaning of agency to what Saba Mahmood has called "a teleology of progressive politics" that celebrates the apparent heroism of the privileged few—namely village leaders, elders, and elites among the poor, overwhelmingly men—who have the resources to engage in formal politics, at the expense of marginalized others, who are excluded or silenced in the dominant narratives of subversion.[91] Disentangling the issues of agency, contention, and inequality through a critical ethnographic stance—and thereby troubling the a priori valorization of resistance on which much of agrarian studies and political ecology scholarship rests—is a key intervention I make in this book.[92]

Ethnography at the Limit: Knowing the Unknown

This book is based on ethnographic research conducted between 2013 and 2016, though in the conclusion I draw on more recent findings from 2018–2020. Before embarking on this study, I had worked in Tanzania as a community development volunteer in 2009 and had been following the EcoEnergy case as a development practitioner between 2010 and 2012, at a time when social movements and NGOs around the world were beginning to mobilize against what became popularly known as the global land grab.[93]

Through deep immersive fieldwork, ethnographers are always intervening in and forming new social relationships, analyzing patterns of connections and disconnections in the worlds they are observing. Yet as with any social practice, relations of power, identity, and positionality shape the doing of ethnography. Being reflexive and transparent about those relations, and the partiality of knowledge produced as a result, is central to the ethics of writing ethnography. It is what gives ethnography its pulse. This section outlines the methodology and methodological considerations that shaped the research process.

In recent years, Bagamoyo has become a highly frequented terrain among researchers, journalists, and activists interested in land politics, as well as business and political elites interested in acquiring and speculating on land.[94] Most visitors to the area have been Europeans, Americans, Indians, or Tanzanians, categories to which I did not belong. In Swahili vocabulary, I was neither a *mzungu* (European/white person), nor *mhindi* (Indian/Asian), nor *mwafrica/mwenyeji* (Black African or "native"), a hierarchical racial order upon which colonialism was built and continues to endure.[95] Neither was I a *mchina* (Chinese), an emergent racial category, though I was sometimes mistaken as one. If villagers didn't call me *dada Mkorea* ("the Korean sister"), they most often called me mzungu, a practice they said was born out of convenience and habit. This process of racialization or racial (mis)reading revealed the hegemony of racial taxonomies in postcolonial Tanzania, and the privileged status associated with my discursive whiteness.[96] People saw me as white by virtue of my appearance, mobility, by and large Western-educated background, and other class signifiers, including a car and the ability to drive. The research process was thus shaped as much by people's sense of confusion and curiosity toward my peculiar whiteness as it was by my own negotiations and interrogations of the complex webs of power relations in which my research participants and I were together entangled.

As race mediated the affective registrars of the ethnographic encounter, gender oriented the spatial dynamics of participant observation. My interactions with women largely occurred in the domestic spaces of homes, kitchens, fields, and gardens. Working within the boundaries of existing gender norms and expectations, I was careful to negotiate my access to the more public spheres men dominated, such as political party meetings, village assemblies, and pubs. I was cautious because my presence and ability to occupy those spaces culturally coded as masculine accentuated not only existing patterns of gender inequality but also the asymmetries of power/knowledge between me and my female research participants. Many women recognized this, too; some were more direct than others in asking me to represent them or bring them news, requests I was happy to oblige to the extent I could. These moments of intersubjective exchange became ethnographic objects in the field, and I tried to capture them in various narrative examples I use throughout the book.

My field site was not a single village or a community but rather the entire investment concession, roughly the size of 20,400 football fields, and the sheer scale of the area and the socioecological diversity within the landscape conditioned my research design. Traveling from the southeasternmost corner of the project site to its northwesternmost corner (where the river meets the railway) required driving roughly forty kilometers on a dirt road over many undulating hills cut by ephemeral and intermittent streams and crossing the Wami River on a dugout canoe (figure 0.6). Over the course of my research, it became evident that there were numerous settlements within the concession beyond the few that official project documents and maps identified (compare figures 0.1 and 0.7). The district government did not formally recognize these settlements as administrative units, nor did it classify all settlers as eligible for compensation as I describe in chapter 3. I use the terms "settlements" and "villages" or "subvillages" throughout the book based on how their residents described their place of dwelling, and as a way to indicate the difference in the status the project afforded them.[97]

In Bagamoyo District, I conducted interviews with 226 people (45 percent female, 55 percent male) from 176 households, the majority of whom resided in the settlements and villages within the concession, and some in neighboring villages.[98] In addition, I conducted eighty-eight interviews with key informants

FIGURE 0.6. Crossing the Wami River by dugout canoe (*mtumbwi*). Author-provided photo, August 2016.

(14 percent female, 86 percent male) in Dar es Salaam and Bagamoyo town. These interlocutors included government officials at the village, ward, district, and national levels; donor agency representatives; corporate executives and employees; and Tanzanian academics, activists, and lawyers with interests in gender, land, development, and agrarian change. The vast majority of these informants were men, highlighting the structural gender inequalities and inequities that shape political participation and decision making.

To assemble the fragmented histories of the landscape, I conducted archival research in the village offices and the Tanzanian National Archives in Dar es Salaam. Owing to the political sensitivity surrounding land issues in Bagamoyo, it was difficult to access archival records from the district land office directly, although a few informants who wished to remain anonymous shared some key maps and documents that informed my analysis. I also collected life histories of local elders as part of the interview process; these individual interviews were complemented by two group oral history and participatory community mapping workshops I organized with approximately twenty elders each.

I combined these ethnographic and historical methods with photovoice and a set of related participatory visual methods. Grounded in critical pedagogy, documentary photography, and feminist standpoint theory, photovoice has gained popularity since its emergence in the early 1990s among scholars and activists

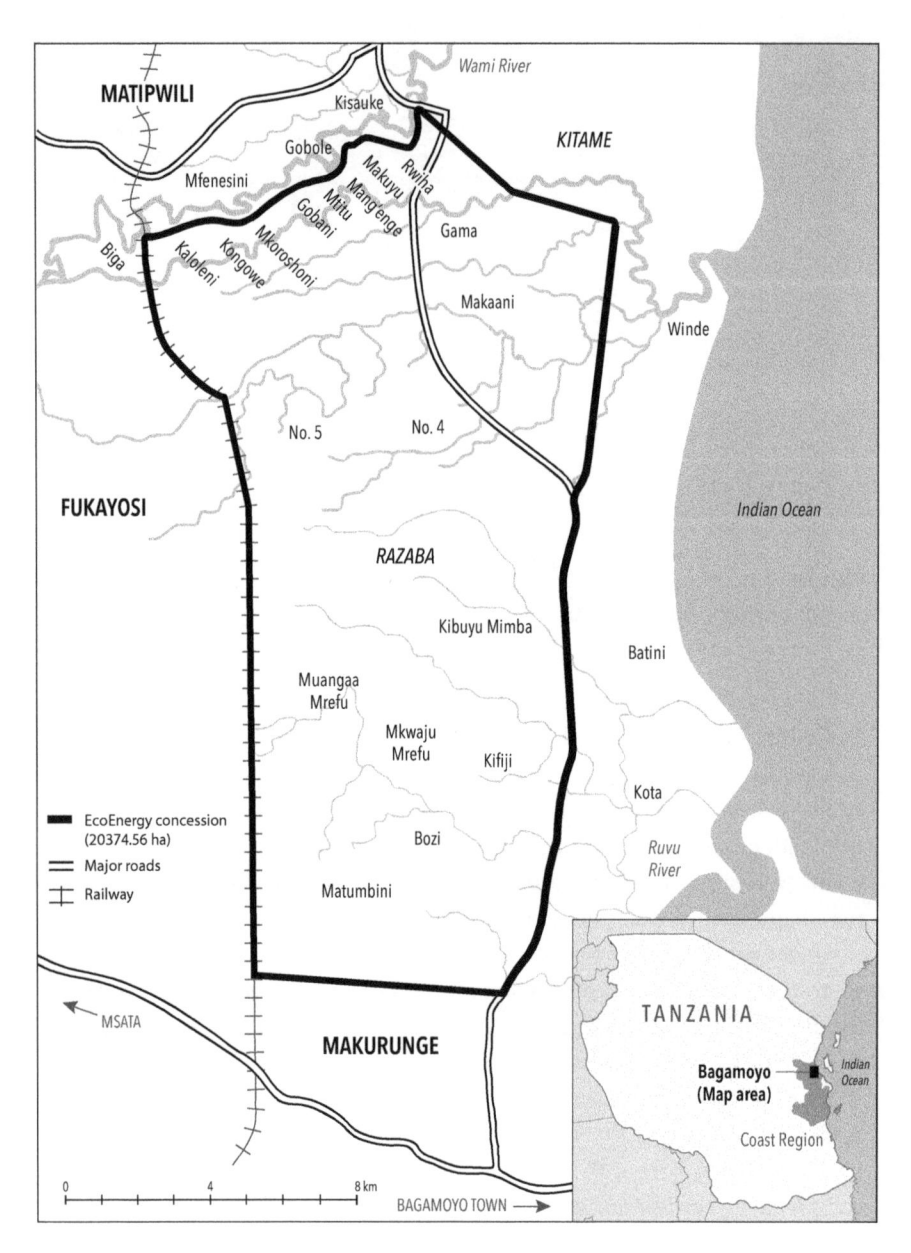

FIGURE 0.7. Map of the EcoEnergy project area situated in the Wami-Ruvu river basin, Bagamoyo District, Tanzania. Map by the author.

seeking a more collaborative approach to knowledge production. Photovoice involves a process in which research participants use cameras to identify and represent particular aspects of their lives, or issues that are important to them and to their communities.[99] For photovoice I worked with eighteen individuals from fourteen households, diverse in their gender, age, marital status, size of landholding, and location within the EcoEnergy concession. After becoming familiar with the camera and the basic techniques and ethics of digital photography, each participant spent on average two months documenting their responses to an open-ended prompt, "What does living on the land with uncertainty look like?" which they were free to interpret the way they saw fit. Of the hundreds of images gathered, I asked each participant to select up to thirty images that were the most meaningful to them; I printed those collections and brought them back for photo elicitation, an interview technique that uses photographs as a guide to facilitate conversations and to elicit specific stories and memories.[100] The participants' photo-narratives and the sense of surprise and intimacy they evoked stimulated new lines of inquiry and shaped the organic flow of my ethnography and the writing of this book.

Chapter 1 begins by tracing the origins and evolution of the EcoEnergy land deal during its formative years between 2005 and 2015. As with numerous other transnational land deals initiated in the 2000s, the EcoEnergy project came into being partly as a result of the convergence of global concerns about climate change, energy insecurity, food shortages, and financial volatility. Under the Kikwete administration, the Tanzanian state also directly enabled it by promoting public policies focused on liberalizing the land market and attracting private investments for large-scale commercial agriculture. While these conjunctures facilitated the making of the land deal, others contributed to its slow unmaking, including financial risks associated with large-scale agricultural ventures, complexities of donor financing, the growing public opposition, and the weakening of the state-investor relationship on the cusp of regime change in 2015.

If chapter 1 situates the EcoEnergy land deal within the political economy of here and now, chapter 2 does so within the uneven historical geographies of enclosure in Bagamoyo and the wider Coast Region. Weaving ethnographic and archival materials, I historicize the bitter landscape in which the sugar project became deeply embedded. The chapter describes people's lived histories and memories of the landscape and makes connections between seemingly disparate settlements and villages in the southern and northern parts of the project site. In so doing, I show how the landscape has been transformed repeatedly through prior inconclusive state-led attempts at dispossession and expropriation of rural resources, including those that are pivotal for cultural rituals, especially for coastal women and girls. These enclosure attempts have included the establishment of cotton and sisal plantations under German (1888–1916) and British colonialism (1916–1961); ujamaa villagization (1968–1973) and the opening of a parastatal cattle ranch (1977–1994) under state socialism; and finally, the creation of the nation's only coastal national park (1999–present) around the same time the area was set aside for the EcoEnergy sugar plantation under postcolonial-postsocialist neoliberalism. The key takeaway of the chapter is that no enclosure, including the latest round, could be permanent in the face of people's persistent struggles to remain on the land and to reoccupy and reclaim what had formerly been their land.

The middle two chapters examine how the land deal was governed on the ground, during the initial planning stages and throughout the prolonged project delay, roughly between the years 2011 and 2016. Chapter 3 takes a close look at one of the earliest planning exercises the government conducted to enumerate local populations and ascertain their eligibility for compensation—an exercise known as the People and Property Count (PPC). By presenting the testimonies of four women of varying social positions and spatial locations, the chapter argues that PPC not only naturalized the patriarchal nuclear family as the basic unit of society but also privileged individual ownership over property. While this discriminated against married women by subsuming their rights to land and compensation for land loss under those of their husbands, it had inadvertently benefited a minority of women "without men," namely widows, divorcées, separated women, and unmarried single mothers. This latter group of women received their own certificates of compensation eligibility by virtue of being situated on the margins of the normalized relations of conjugality and familyhood. And thus they were determined to protect what they perceived as a rare advantage that happened upon them, although their sense of entitlement was inherently limiting in that the PPC was ultimately designed to facilitate and legalize dispossession.

Chapter 4 examines the role of an unlikely pair of actors whom EcoEnergy contracted to manage the day-to-day governance of the land deal in Bagamoyo: a foreign development consultant and district paramilitary forces. I argue that these actors' divergent interests and identities, as well as the different conjunctures at which they were deployed on the ground amid the project delay, gave rise to two contradistinctive mechanisms of governance. The first was the foreign consultant's biopolitical intervention, which focused on improving the lives of to-be-displaced populations through various nonagricultural skills training courses. The second was the district paramilitary forces' necropolitical intervention, which focused on land use restrictions, surveillance, intimidation, and physical violence. The chapter argues that both enactments of power, like the

PPC, not only laid the foundation for rural dispossession, but also reinforced normative expectations of gender: what activities and behaviors are considered natural or respectable ways of being women and men, feminine and masculine in rural Tanzania.

The last two chapters explore the range of contentious politics and grassroots resistance. Chapter 5 focuses on the quotidian, covert, and semiorganized means through which different individuals and groups responded to the constraints the project imposed on their everyday life. People engaged in ordinary speech acts, such as rumor and gossip, to make sense of their liminal collectivity and to build social bonds. To ensure their survival and sustenance in the face of continued mgambo violence, those with fewer fallback options, particularly women "without men" and migrant youth, engaged in more illicit and risky practices, such as guerrilla-style farming, unauthorized charcoal production, and unlicensed alcohol brewing. On the other hand, a small group of young, educated, activist-minded migrant men engaged in political organizing; they started petitions, wrote letters to politicians, and involved journalists to demand their rights as agrarian citizens. Yet the group's membership remained exclusionary toward women, elders, and longtime residents, which limited the extent and effectiveness of their work. Although these varied forms of everyday resistance were, at times, contradictory and conflictual and did not contribute directly to the suspension of the land deal, they nonetheless highlight the heterogeneous expressions of subaltern agency.

Chapter 6 explores another, more privileged political strategy of refusal: litigation. It focuses on a trespass lawsuit three male elders filed against the Tanzanian government and EcoEnergy. While the plaintiffs justified their action as rightful resistance to land grabbing, the High Court of Tanzania dismissed their case with costs. Triangulating from observations, interviews, and court documents, the chapter argues that the elders' apparent resistance is more appropriately understood as gendered "lawfare" from below. I describe how the elders drew on their multiple positions of privilege to exclude a diverse array of legitimate resource users, including, most immediately, their wives. Their recourse to law, thus, was perverse both in its design and its social effects; not only did it increase the uncertainties surrounding the rights and status of local residents vis-à-vis the plaintiffs, but it also exacerbated existing local inequalities across gender, class, generation, and social status.

In the conclusion, I return to the Tanzanian government's decision to terminate its relationship with EcoEnergy, which led to the investor filing an arbitration claim at the World Bank's International Center for Settlements of Investment Disputes in 2017. While the case remained pending, the Tanzanian government reallocated part of the land to the largest domestic conglomerate, again for sugarcane production. The conclusion offers an update on this "new" land deal and examines the implications and stakes of EcoEnergy's arbitration case. It also reflects on the relevance of the book's findings for ongoing political debates on development and nation building in Tanzania, as well as global policy efforts to promote ostensibly responsible and gender-sensitive investments through voluntary codes of conduct for investors and states.

1
THE MAKING OF A SWEET DEAL

> A study of global connections shows the grip of encounter: friction.
>
> Anna Tsing, *Friction: An Ethnography of Global Connection*

To say the EcoEnergy land deal was ambitious would be an understatement. It promised to deliver many firsts for Tanzania: the nation's first new sugarcane plantation in over four decades, its first biofuel project, and the first transnational public-private partnership (PPP) that would shape the future of rural development in the years to come. The land deal was also anticipated to become one of the largest agricultural investments in East Africa to date, with a total estimated cost of USD 500 million.[1] Yet, despite millions of dollars invested and over ten years of planning, capital accumulation for EcoEnergy remained a far cry from reality.

This chapter analyzes the myriad actors, processes, and relationships through which the EcoEnergy land deal came about and unfolded in unanticipated ways during its formative years, between 2005 and 2015. Focusing on this critical period, the chapter reveals how the various political, economic, social, environmental, and legal challenges the company faced—and the opportunities these problems created—contributed to the making and remaking of the land deal. What began as a purely private investment that would produce sugarcane biofuels for export to Europe to fight global climate change later became a PPP that would supply sugar for the Tanzanian domestic market to promote food security and agricultural development. Modernist ideas of progress and property as well as the flexibility of sugarcane as a commodity with multiple, interchangeable uses underlay both iterations of the project. Although the nature of the project evolved over time, it continuously drew on a "win-win" rhetoric that effectively masked the structural power asymmetries between the actors differentially implicated in and impacted by the land deal. Keeping the project alive and maintaining a semblance of stability and legitimacy thus required a tremendous amount of work—political, material, and discursive—especially on the part of the corporate investor. Yet, as I go on to show, these continual efforts to fix and sustain the project were not durable nor impervious to contestation; civil society would expose their limits at every turn. And so would, perhaps to EcoEnergy's greatest disappointment, the Tanzanian state.[2]

Banking on the Climate Crisis: The Elusive Promise of Biofuels in Africa

The beginnings of the EcoEnergy land deal trace to the global biofuels boom and to a Swedish company, known as Sekab (Svensk Etanol Kemi AB). Established in 1985, Sekab is one of the largest producers and distributors of biofuels and biochemical products in Europe, with an annual turnover of over USD 200 million. The company is 70 percent owned by three public utility energy companies in northern Sweden and 30 percent by a private company, EcoDevelopment in Europe AB (hereafter EcoDevelopment).[3] Little public information exists on EcoDevelopment, but an EcoEnergy report describes the company as being "owned by 18 Swedish citizens and business leaders, representing entrepreneurial as well as industrial and financial expertise."[4] As I will return to later, prominent shareholders of EcoDevelopment include Per Carstedt, the former Sekab CEO turned EcoEnergy executive chairman; and his brother Göran Carstedt, a former executive of Volvo and IKEA in Europe and North America, who also served as the senior director of the Clinton Climate Initiative in 2007–2008.[5]

After enjoying a first boom during the oil crisis of the 1970s, the global biofuel industry experienced another major boom in the early days of the new millennium, when governments around the world sought to diversify their energy supply and to reduce greenhouse gas emissions to mitigate climate change. In Europe, national climate policies and EU directives in the early 2000s promoted the production and use of biofuels and other renewable energy in the transport sector through incentives such as consumption mandates, subsidies, tax credits, federal funding for research and development, and pro-ethanol trade agreements.[6] Since the company's early days, Sekab had been investing in the production of second-generation biofuels from lignocellulosic biomass such as grass, sawdust, and forest residues.[7] However, with the slow development of cellulose-based refinery technologies and high production costs at a time of increasing demand, the company decided to outsource cheaper first-generation biofuels derived from sugars, grains, and oilseeds. Like many other investors at the time, Sekab looked not only to existing suppliers in Brazil but also to starting up new "greenfield" projects in countries like Ghana, Mozambique, and Tanzania. The company saw Africa as the next frontier for biofuels, a laboratory in which new spatial fixes to the climate crisis would be tested.[8]

In a presentation at a 2008 International Energy Agency workshop on "Biofuels for Transport," Per Carstedt outlined the reasons he believed biofuel investments were highly pertinent to and beneficial for Africa. First, he argued that the continent was vulnerable to crude oil price volatility, and thus it had much to gain from the domestic production, consumption, and export of renewable energy. Further, he believed that Africa was endowed with the "best natural conditions" for large-scale irrigated agriculture for feedstock production.[9] It had "surplus land and water," an "available labor force," and a "gigantic need for social and economic development."[10] Taking the example of Tanzania, Carstedt boldly claimed that 1 to 2 percent of the nation's arable land was all that was needed to provide 100 percent of its domestic transport fuel needs.[11]

Carstedt's determination to act on climate change may well have informed his decision to invest in biofuel production overseas. Yet the Swedish informants I interviewed were skeptical, with some suggesting that his motivations were likely linked to his private business interests at home in the energy and transport sectors. One informant described him as "a key figure in the motor vehicle industry in Sweden."[12] Another characterized him more crudely as a "salesman with personal interests in selling more cars," especially flex-fuel vehicles that ran on ethanol.[13] Public records, indeed, indicate that Per Carstedt is the CEO of Carstedts Bil AB, an authorized Ford dealer in northern Sweden.[14]

Sekab first knocked on Tanzania's doors in late 2005. At the time, there was not a single biofuel project in the nation, nor were there any policies, laws, or institutional mechanisms for governing such operations. In November 2005, at a national biofuels stakeholders meeting organized by the Ministry of Energy and Minerals, Sekab presented a proposal for a joint-venture biofuel project with a Swedish organization, the BioAlcohol Fuel Foundation, and a private Tanzanian firm, Community Finance Company.[15] This meeting was part of a series of events the ministry had organized in response to the findings of a report by GTZ (German Technical Cooperation Agency) titled *Liquid Biofuels for Transportation in Tanzania*. Released in September 2005, the report was the first-ever publication to explore the possibilities of biofuels development in the nation. The most notable recommendation of the report was that the Tanzanian government should move fast to take advantage of new market trends and growing investor interests in biofuels. The report stated, "In order to quickly proceed with the introduction of biofuels in Tanzania, the Government should take immediate action to enter the learning-by-doing process—and not wait for results and policy advice from the [yet-to-be-established National Biofuel] Task Force."[16] The study made no mention of the need for public debate on the issue, except that "the [Tanzanian] population has to be informed about the significant benefits and opportunities offered by biofuels as alternative transport fuel."[17] In October 2005, the World Bank released a similar report, titled *Potential for Biofuels for Transport in Developing Countries*, in which it highlighted sugarcane ethanol as "likely to offer the best chance of commercial viability" in the near future for low-income countries like Tanzania.[18] In assuming biofuels as an economic development opportunity and a "clean" alternative to fossil fuels, neither report gave serious consideration to their potential adverse socio-environmental impacts.

In March 2006, the government, via the Ministry of Energy and Minerals, established the National Biofuel Task Force, comprising eleven public-sector and two private-sector representatives.[19] The role of the task force was to develop guidelines that would provide investors with a minimum requirement necessary to ensure so-called sustainable biofuel production in the country. In June 2006, however, long before the task force could begin work on the guidelines, the government, via the Ministry of Planning, Economy, and Empowerment, signed a memorandum of understanding (MoU) with Sekab and its proposed joint venture partners to "kick-start the development of a long term and sustainable BioEnergy platform in Tanzania."[20] The government embraced the GTZ's advice on adopting a rapid, experimental approach to biofuel investments, even in the absence of regulatory frameworks and robust evidence of such projects' impacts on society and the environment. Several months following the MoU signing, Sekab established Sekab BioEnergy Tanzania (hereafter Sekab BT) as its local subsidiary, with Per Carstedt serving as executive chairman.[21]

It was only in late 2008 that the task force released a first draft of the "Guidelines for Sustainable Liquid Biofuels Development in Tanzania." With the Norwegian and Swedish development agencies providing USD 3 million in financial aid, the task force finalized the document in November 2010.[22] However, activists and researchers in both Tanzania and Sweden heavily criticized the guidelines. International nongovernmental organizations (NGOs), such as ActionAid and the World Wildlife Fund (WWF), argued that the task force had overlooked the negative impacts biofuel production could have on local land rights, food security, and biodiversity.[23] They argued that the task force, which was led by the Ministry of Energy and Minerals, saw biofuels primarily as an energy issue, without attending to their interconnections with land, agriculture, and rural livelihoods.[24] Writing about Sekab BT's proposed project in particular, the WWF argued that there were major discrepancies between the company's stated objectives and the perception of local communities. It also criticized the company's claims to "sustainability" or "carbon neutrality," because biofuel projects, with their large-scale, intensive land use and land use change, could end up emitting more carbon dioxide than they would be able to sequester.[25] Another fundamental problem with the guidelines was that they were voluntary and only a stopgap measure rather than legally enforceable policies.

Criticisms against Sekab BT and biofuel investments in Africa more broadly would redouble in subsequent years, in the context of growing public awareness and debates on global land grabbing—a topic to which I will return later. To better contextualize this forthcoming discussion, the next section describes the steps Sekab BT took to acquire land in Tanzania and the problems and contradictions that arose in the process.

The Politics of the Land Question

According to a senior official I interviewed in the Prime Minister's Office (PMO), many of his government colleagues were initially impressed with Sekab's investment idea. "People thought it was so sexy, so nice, [and] so powerful."[26] But there were also others who were doubtful or grew wary over time of the investors' unrealistic demand for land, the lack of regulatory oversight, and the risks involved in large-scale investments, especially in agriculture. The PMO official himself was most concerned about the land question—whether and how the government would alienate land from existing users:

> I told many people this. I said, "I am sorry, but without land reform, there will be no project, no biofuels." Let me start from the whole of Africa. From Dar es Salaam to Congo, or from Cairo to Cape Town, there isn't a square meter of land that does not have a claim on it. Someone is using it or occupying it for something. I've said it in many conferences including at the World Bank's Land and Poverty Conference in Washington, DC. People tell me I am just being funny or stupid, but I am serious when I say this: From Cairo to Cape Town, there's a claim on every square meter of land. Land is politics.[27]

Land acquisition in Tanzania, including land transfers to foreigners, is guided by various land laws and policies, including the Land Acquisition Act of 1967, the National Land Policy of 1995, and the Land Act and Village Land Act of 1999. All land is state-owned, but the administration of public land is decentralized. The Land Act of 1999 classifies public land into three categories: village, reserved, and general land.[28] Village land, the vast majority of rural land in the nation, is under the jurisdiction of village councils; reserved land, which includes protected and hazardous areas, is governed by designated statutory bodies; and general land, comprising all remaining land and what the law ambiguously calls "unoccupied and unused village land," falls under the jurisdiction of the commissioner for lands.[29] All three land categories, at the end of the day, are governed by the commissioner, who is answerable to the minister of lands, housing, and human settlements. To become investible, all land categories must be converted to the general land category.[30]

Investors interested in acquiring land are first expected to contact the Tanzania Investment Center (TIC), an investment promotion agency or a "one-stop shop" for investors established by the Tanzania Investment Act of 1997.[31] For land requests below 250 hectares (just under 1 square mile), investors can negotiate directly with villagers upon formal introduction by the TIC and local and regional authorities.[32] For areas larger than 250 hectares, the minister of lands has the authority to approve or refuse the land transfer, upon considering recommendations by village and district councils.[33] Ultimately, however, the president has the prerogative to expropriate and transfer any area of village land, including those that have been occupied for generations, to general land for investments of "public interest."[34] What counts as "public interest" is undefined in the law and thus open to contestation.[35] The TIC is responsible for maintaining a "land bank," an inventory of general land parcels available for transfer to foreign investors under derivative rights of occupancy. According to a TIC official, however, the land bank has been inoperative, in part because of the insufficient amount and scattered nature of surveyed land parcels, as well as the general lack of resources to maintain the system's day-to-day functioning.[36]

When I asked about the status of the land bank, an assistant commissioner for lands was unequivocal about its nonexistence. He shared a concern similar to the one the PMO official expressed earlier. The land bank did not exist, the assistant commissioner argued, because there was no such thing as "unused" or "free" land in Tanzania:

> We don't have a land bank. Whoever told you that we have a land bank in Tanzania is not telling the truth. We are struggling to create one. Let me tell you: The Land Act observes that each investor who comes to Tanzania should get land from the land bank. There was an attempt since 2000 to establish a land bank. The problem is that, in Tanzania, we don't have such a thing as a "no-man's-land." We don't have land that is "free." So even if you identify village lands you want to put in the land bank, you must first pay compensation to those villagers. But we don't have money to pay compensation. That is the problem. That's why up to now we don't have a land bank. That is the reality. Only 2 percent [of all public land] is general land, so we need to make more

land available for investors from the villages. But to establish a land bank we need to establish a land compensation fund as provided by the Land Act of 1999. But we still don't have the fund right now. So in that case, if an investor comes to Tanzania and wants land, there are a few options: (1) Go to the market him- or herself to look for land, or (2) in certain circumstances, if there is a big project like Sekab [later EcoEnergy], then the government can initiate the process of finding land for the investor.[37]

In the absence of a land bank, Sekab indeed went about its own way to obtain land in Tanzania. An early project document indicates that the land acquisition process began in mid-2006 in Sweden through a survey of "satellite images for the 200 km boundary along the coast of Tanzania."[38] Based on these images, the company tentatively identified Bagamoyo and Rufiji, two of the largest districts in the Coast (Pwani) Region, as having the greatest potential for large-scale irrigated sugarcane production. In February 2007, Sekab BT contacted the district commissioners in Bagamoyo and Rufiji to inquire about land availability. The company then reached out to the TIC to formally request land in these districts; the TIC subsequently asked the district executive directors to assist the company with the land acquisition process.

In late March 2007, Sekab BT wrote to the Bagamoyo District executive director, requesting access to a former state cattle ranch, known as RAZABA (Ranchi ya Zanzibar Bagamoyo). According to the company executive Per Carstedt, the decision to site the project in that specific location was not his but the central government's. He had initially hoped to develop a cluster of projects in Rufiji District involving at least two hundred thousand hectares, but he said preparing for an investment of that size would require many years of feasibility studies as well as the displacement of numerous villagers within the Rufiji river basin.[39] When he raised this issue with the Tanzanian government, he was allegedly directed to an "idle" state ranch in Bagamoyo where he could, arguably, start the project relatively quickly:

> We told the Tanzanian government that this investment in Rufiji is complex; it will take four to five years [to set up]. Then the government said, "We need development. Can't you cut some corners? Can't you do it quicker?" But we said, "No, this is how much time it would take to do the project in a proper way." So then the government came back and said, "Look, here, we have this government-owned land [in Bagamoyo], right? It's idle; it's a defunct cattle ranch, and you don't need to do a river basin study." . . . So that's how the project came about. That's how we ended up in Bagamoyo. It was not our initiative.[40]

I interrogate the issue of "idle" land in chapter 2, but for now it is important to note that Bagamoyo is the birthplace and political home of the then-incumbent president, Jakaya Kikwete. Several informants, including a Tanzanian Sekab BT employee, argued that Kikwete's decision to direct the investor to Bagamoyo was politically expedient: "You know, presidents, ministers, and other high-level officials in this country want to see their names reflected in places where they come from. Kikwete had been an MP of Bagamoyo for a long time, but its economy is still struggling, and the living standards of its people are still very low. Kikwete knew that the district was in desperate need of investment."[41]

With regard to the company's request for the ranch, however, the Bagamoyo district executive director responded quickly, stating that the district did not have authority over the property, as it belonged to the government of Zanzibar, the Indian Ocean archipelago whose reunion with Tanganyika created the United Republic of Tanzania in 1964. After five months of impasse, Sekab BT wrote directly to the State House in August 2007 to resolve the issue directly with the president. Citing the Land Act of 1999, which vests all land in the head of state, the company appealed to President Kikwete to expedite the land acquisition process. Three months later, on November 4, 2007, the Ministry of Planning, with whom Sekab had signed an MoU a year prior, notified the company of the government's decision to allocate them a significant portion of the RAZABA ranch. The company would have access to roughly twenty-two thousand hectares, and the Zanzibari government would retain the remaining six thousand or so hectares. The exact size of the concession would be confirmed once the Ministry of Lands conducted a land survey, the costs of which Sekab BT agreed to cover. The ministry completed the land survey in January 2008, and in May of that year the State House chief secretary confirmed the subdivision of the ranch and an allocation of 20,570 hectares to Sekab BT; a formal title was to follow suit, though as I discuss below, it would not be completed until five years later, in 2013. The land concession area ended up being much smaller than what the company had hoped to develop in Rufiji, but Carstedt said he accepted the government's offer, thinking that the Bagamoyo project could serve as a "starter" or "model" farm for a larger project in the future.[42]

To summarize, the legality of Sekab BT's land acquisition process is murky and debatable. The company's quest for land was anything but straightforward; it involved conversations with numerous government authorities at the district and national levels. Without a land bank at the TIC, the identification and transfer of land remained ad hoc and politically facilitated by the Kikwete administration and the president himself. During this process, however, the so-called defunct status of the state cattle ranch was never questioned. The government, and by association Sekab BT, arguably made an a priori assumption that the ranch was "unoccupied and unused" and thus already in the general land category, ready for investment. Consequently, no residents were consulted or given the opportunity to debate, consent to, or refuse the proposed investment, the ramifications of which are the subjects of later chapters.

From Sekab to EcoEnergy: Becoming a Development Project

While the land transfer was being formalized, Sekab was beginning to face numerous pressures, internal and external, that threatened its own legitimacy and viability. The following sections highlight the oppositions and obstacles, as well as the opportunities, Sekab BT encountered between 2009 and 2010, and how they ultimately forced the company to sell, change its name, and revise its business strategy.

Crises of Legitimacy

As Sekab BT increasingly became a target of activism against global land grabbing, the company also became subject to intense public scrutiny at home in Sweden.[43] First, Sekab and the three municipal energy companies that owned majority shares in the company were accused of failing to inform local taxpayers about the firm's activities in Africa. In September 2010, an aggrieved resident of Örnsköldsvik, a municipality in which Sekab is headquartered, filed a complaint at a local administrative court, and the court eventually ruled that it was illegal for the municipality to finance Sekab's operations in Africa, let alone establish subsidiaries overseas.[44] Prior to this ruling, controversies had been brewing over Sekab BT's claims of sustainability. In early 2009, an Oslo-based independent journal, *Development Today*, exposed how the company had tampered with its environmental and social impact assessment (ESIA) to portray the biofuel project "in the best light."[45] The original ESIA, for instance, had expressed serious concerns about the project's significant demands for water, the loss of biodiversity and common resources, and conflicts with local communities and an adjacent coastal national park (chapter 2). All these concerns were either deleted or toned down in the final report submitted to and approved by the Tanzania National Environmental Management Council in April 2009.[46] An independent Swedish consultant who had drafted an earlier version of the ESIA also claimed that Sekab BT had forged her signature in the final document, which further tarnished the company's integrity and reputation.[47]

Furthermore, the growing public skepticism around biofuels as greener alternatives to fossil fuels and the slowdown in the growth of biofuel demand in the post-2009 period called into question the relevance of Sekab BT's proposed project. As a Dar es Salaam–based official with Sida, the Swedish International Development Cooperation Agency, explained in an interview,

> After the financial crisis, food price crisis, suddenly there was no real interest anymore in large-scale ethanol production. Ethical issues also came up on why a foreign corporation would use arable land in Africa to produce something other than food crops when the majority of African populations are food insecure. Plus, studies showed that large-scale monoculture production of feedstock for ethanol was not using any less fossil fuel than actual drilling for crude oil. Ethanol, at this point, became no longer attractive as a solution to address climate change. And especially for Sekab BT, it's been really tough for them to get the financing together.[48]

Indeed, the financial crisis did not leave Sekab unscathed. Between 2005 and 2009, the company reportedly incurred losses of SEK 170 million (USD 20 million) in Africa alone.[49] Faced with the difficulty of raising capital following the 2008 global financial crisis, Sekab BT approached Sida in July 2009 to request a credit enhancement guarantee. Guarantees as a form of development aid have grown in relevance only in recent years, most notably in the energy and extractives sectors, as a response to decreasing volumes of traditional overseas development assistance and to growing private-sector interest in international development. A credit enhancement guarantee from reputable development agencies like Sida could allow companies like Sekab BT to share or transfer risks, to borrow funds from local banks on better terms, and to "crowd in" other funding sources.[50]

At the time of its application to Sida, however, Sekab BT was in the process of being sold. In a letter to Sida, the company's managing director Anders Bergfors explained the company's decision to sell as a way to "separate the African ventures from the municipalities in northern Sweden as well as to maintain the Swedish connection."[51] In the same letter, he indicated that the company was in the process of inviting the Tanzanian government to become a 10 percent shareholder in the new project to promote the nation's first "Public Private Partnership in the AgroEnergy [sic] sector."[52] On October 21, 2009, EcoDevelopment (the private, minority owner of Sekab, headed by Per Carstedt) bought Sekab BT for a song, at just SEK 400 (USD 50).[53] About a week after this acquisition, Sida formally rejected the company's application for the credit enhancement guarantee.

Based on a review conducted by Sida Helpdesk for Environmental Assessment, the agency denied Sekab BT's request for several reasons. First, the agency could not provide guarantees for the sole purpose of financing a firm's commercial development costs; second, the company failed to specify where, how, and for what it intended to use the funds; third, the agency believed that the project's environmental and social costs would outweigh its purported benefits; and finally, the agency deemed the legal frameworks in Tanzania "too fragile" to govern biofuel investments of the type and scale Sekab BT proposed.[54] In brief, Sida determined that there were too many risks associated with the project and that the company lacked the "social license to operate," as the official I interviewed put it.[55]

Once sold to EcoDevelopment, Sekab BT was renamed Agro EcoEnergy Tanzania. The ownership structure of this new company is complex; "impervious" was how an internal Sida document described it.[56] In the process of acquiring Sekab BT, EcoDevelopment established a new parent company in late 2010 called EcoEnergy Africa AB, registered in both Sweden and the tax haven of Mauritius.[57] As of 2014, EcoEnergy Africa AB was 99.8 percent owned by EcoDevelopment and 0.1 percent each by Anders Bergfors and Arvind Puri, the managing director and the chief financial officer of Agro EcoEnergy Tanzania (and previously Sekab BT).[58] In late 2010, Agro EcoEnergy Tanzania also established a special-purpose-project company, called Bagamoyo EcoEnergy Limited, owned 99 percent by Agro EcoEnergy Tanzania and 1 percent by EcoEnergy Africa AB.[59] What I refer to as EcoEnergy in this book for simplicity and brevity includes all corporate entities who hold shares in one way or another in the sugarcane venture in Bagamoyo.

A Sweet Salvation

Based on conversations with EcoEnergy executives and employees, the company's decision to continue operating in Tanzania, despite the collapse of Sekab BT, had much to do with the political economy of sugar in Tanzania. Just as President Kikwete had influenced the siting of the project in Bagamoyo, this time he had allegedly asked the company to stay and help alleviate the nation's sugar deficit, which amounted to about three hundred thousand tons per year. As a Tanzanian EcoEnergy employee explained, "When Sekab BT was about to close shop and leave the country, the company executives went to visit President Kikwete to thank him for the opportunity, etc. But Kikwete said, 'No, you can't leave. We have sugar shortage in this country. Go back to your drawing table.' So that's how EcoEnergy Sugar Project was born."[60]

Under the assumption that the production capacity of the four existing sugar mills in the country would soon peak and that the annual domestic demand for sugar would increase to over one million tons by 2020, Kikwete launched a series of new agricultural initiatives between 2009 and 2013 to improve productivity at existing mills and attract investments in new sugar plantations and factories.[61] Under these initiatives, which I delve into later, Tanzania planned to produce as much as 4.4 million tons of sugar per annum by 2030, compared to the annual production of 300,000 tons in 2012.[62] Prioritizing domestic production was expected to result in large foreign exchange savings for the national state. To address the sugar deficit, the government had hitherto waived or reduced taxes on imported sugar from countries such as Thailand, India, Brazil, and Indonesia.[63] A senior official in the Ministry of Agriculture, Food Security, and Cooperatives explained that the government "basically chose sugarcane as import substitution [strategy]" and to achieve economies of scale: "With sugar, you can have a large farm, you can mechanize. It's easier to move [to a large-scale production system] with sugar than coffee, for instance, which is almost entirely smallholder based. Because the government wanted *big* results *now*, it meant we didn't have time to go into details of engaging with smallholders and so forth. But with sugar, we could have a large farm and then have the smallholders become outgrowers."[64]

Ensuring a steady and cheap supply of sugar for domestic consumers was also arguably necessary if ruling political elites were to maintain their popularity and legitimacy. As the PMO official introduced earlier in this chapter put it, "If citizens aren't able to have tea with sugar in the morning, they will not vote for anyone. If there is no sugar, no cheap sugar in the market, people will riot!"[65] For Per Carstedt, investing in sugar production for the Tanzanian domestic market made both political and business sense. Consider his long response below to my question as to why he decided to stay in Tanzania, and note in particular the shift in his investment rationale: from producing biofuels to combat global climate change to helping Tanzania achieve national economic development. Also evident in his response is a modernist concern over the so-called nutrition transition, an idea that as population grows and national incomes rise, people will consume more and more sugar along with other nutrients like salt and fat:

> We did some calculations together with a number of other institutions and using the UN projections for population growth for the next twenty years, we made assumptions on how the consumption per capita of sugar would change. Economic growth [for Tanzania] is 6 to 7 percent annually, and the consumption of Coke, consumption of things like that is going to increase 10 percent a year. A huge part of the population will consume sugar, all right? They will consume some basic [substances] like sugar and salt, which they did not have before. The population of Tanzania will grow from about fifty-three million today to ninety-four million by 2035, and you can assume that the consumption of sugar will grow, too. So if you take those two assumptions—growth of population

and consumption of sugar per capita—Tanzania will triple its sugar consumption in the next twenty years. Still, the consumption per capita here will be less than half that of Europe, and less than one-third of that in the US.

If you look at the current level of production, domestic producers produce roughly 300,000 tons, but by 2035, people will consume 1.8 million tons. So if you look at that opportunity, we can say, well if you want to sustain your people, you need to increase domestic production. The alternative is, if you don't get your act together, you will not be able to produce, so you'll have to import. And if you take the average price of sugar over the last five years and multiply that by the volume that needs to be imported over the next twenty years, that would be USD 10 billion.... If you are talking about development, you cannot but cash in on the low-hanging fruits. I mean, aiming to do iPads, iPhones, cars, or machines would probably be too high up in the tree, but there are low-hanging fruits like agriculture, particularly if you consider that the nation has favorable natural conditions, like climate, water, and land. And there is a need to provide jobs in rural areas. That's probably the most dramatic thing. If you don't do that, you will have social crisis. It's a ticking bomb; [if the government does not act,] the youth and the opposition will go to the streets.[66]

This change in investment priority from biofuel to sugar production was conditioned by not only the shifting political-economic processes at the national level, but also the material and discursive flexibility of sugarcane as a commodity. As recent studies of land grabbing have shown, "flex crops," such as sugarcane, corn, soy, and oil palm with multiple and interchangeable uses (e.g., food, fuel, feed, and industrial products), have allowed investors to strategically craft and switch between different legitimating narratives for their projects.[67] Carstedt's quote above and the following two excerpts from Sekab BT's and EcoEnergy's project documents from 2008 and 2017 respectively illustrate the degree to which sugarcane can be and has been "flexed" to meet the company's changing needs.

Sekab BT, 2008: The proposed project will establish the first large scale renewable bioenergy project in Tanzania, which will demonstrate the valuable contribution Africa can make towards the global climate challenge.[68]

EcoEnergy, 2017: Tanzania is currently importing more than 50 percent of sugar consumed.... If domestic production is not allowed to be developed, the volume of sugar to be imported will further increase in the coming decades due to strong population growth.[69]

The transformation of Sekab BT to EcoEnergy also entailed a reconfiguration of the company's relationship with the Tanzanian state and the local communities. EcoEnergy was no longer simply a product of private investment but rather a PPP for agricultural and rural development. To that end, EcoEnergy proposed to implement an outgrower and community development program, which would support from fifteen hundred to two thousand local smallholder farmers living within a forty-kilometer radius of the sugar mill to supply cane on a contractual basis.[70] To participate in the program, however, smallholder farmers were expected to form their own companies, consolidate their land into approximately one-hundred-hectare irrigated block farms, and take out a minimum of USD 1 million in loans—a sum equivalent to a thousand times Tanzania's GDP per capita—to finance their farms.[71] From EcoEnergy's perspective, the project—inclusive of the plantation, mill, and the outgrower scheme—was a "win-win"; it would benefit not only the investor and the state, but also local communities in "long-term value creation."[72] A British (male) engineer and outgrower specialist whom EcoEnergy contracted described the integrated plantation-outgrower model as a "PPPP: people-public-private partnership." The engineer's quote below, which is redolent with paternalistic and racist undertones, epitomizes the continuing salience of modernization theory in guiding global development:

> The outgrower [program] is about empowering people to shift from a "survival mode" to a "grow mode." In survival mode, farmers are not used to planning; they are concerned about the day-to-day; they are not winning; people are avoiding risk. In grow mode, it's like African farmers are finally coming out of their caves and joining the modern world, a world of commercial agriculture. In grow mode, they are winning, thinking long-term, they are not avoiding, but *managing* risk. It is very rare that a farmer will make enough profit to pay for all operation costs, so they need to go to the bank. And banks love sugar outgrowers because they can use the cane supply agreement with the mill as a collateral. And then the mill can pay the bank before paying the farmer through a stop-order [salary deduction]. This is how the outgrower becomes part of the modern world. We are trying to use sugarcane to *grow* people.[73]

Tanzania's Road to a Green Revolution

Before discussing EcoEnergy's trajectory further, it is important to pause and situate the "new" sugar project in relation to the political economy of agricultural development in Tanzania. As I alluded to earlier, during the Kikwete administration (2005–2015), national agricultural policies shifted from a model based largely on state and donor efforts to improve smallholder agriculture to one that prioritized the role of the private sector and PPPs to promote large-scale agro-industrial operations. When Sekab first arrived in Tanzania, the country had been facing continued economic stagnation and growing public debt, amounting to nearly half the national gross domestic product in 2004/5.[74] After two decades of neoliberal reforms imposed by international financial institutions (IFIs), Tanzania still relied on foreign aid to supply 50 percent of its total national budget.[75] Politically, the national state also confronted a legitimation crisis over its inability to deliver material welfare benefits to its citizens, the vast majority of whom resided in rural areas and relied on agriculture for their primary livelihoods.

Against this backdrop, Jakaya Kikwete was elected the fourth president of Tanzania in December 2005, with an overwhelming 80 percent of the popular vote. Following his campaign slogan, "*Ari mpya, nguvu mpya, kasi mpya*" (New zeal, new strength, new speed), Kikwete launched several new development initiatives building on those promoted by his predecessor, the late Benjamin Mkapa, under whom Kikwete served as minister of foreign affairs. To expedite the agricultural sector development strategy that the Mkapa administration had formulated in 2001 to quality for debt relief from the IFIs, Kikwete launched the Agricultural Sector Development Program (ASDP) in 2006.[76] As a state-driven initiative in line with the National Strategy for Growth and the Reduction of Poverty, as well as the African Union's Comprehensive Africa Agriculture Development Programme, which aimed to increase public spending in agriculture, the ASDP quickly gained the support of donor agencies, including the World Bank, the African Development Bank (AfDB), and the UN's International Fund for Agricultural Development (IFAD).[77]

The implementation of ASDP, however, remained slow. This was due not only to tensions between the Tanzanian government and donor agencies on investment priorities and the role of the private sector, but also to significant drops in foreign aid volumes following the global financial crisis.[78] The post-2008 conjuncture thus marked a major shift in the Tanzanian government's as well as donors' approaches to financing agricultural development. In its *World Development Report 2008: Agriculture for Development*, the World Bank's first flagship report in twenty-five years to focus exclusively on agriculture, the bank stressed the need for national governments to stimulate private investments to improve agricultural productivity and to engage strategically in PPPs to boost competitiveness in the agribusiness sector.[79]

This emphasis on privatization of agricultural development became a cornerstone of Kikwete's *Kilimo Kwanza* (Agriculture First) initiative, launched in

August 2009. Kilimo Kwanza was not a policy but rather a two-page resolution that outlined the nation's vision for a "Green Revolution to transform its agriculture into a modern and commercial sector."[80] Whereas the ASDP had prioritized the provision of public goods to improve smallholder production systems, Kilimo Kwanza explicitly called on the private sector to "substantially increase its investment and shoulder its rightful role" in the implementation of agricultural modernization in Tanzania.[81] Kilimo Kwanza in many respects mirrored the discourse of the Alliance for a Green Revolution in Africa, which the Rockefeller and the Bill and Melinda Gates Foundations formed in 2006 and opened an office in Tanzania in 2007 to replicate the earlier agricultural modernization experiments in Latin America and Asia. Kilimo Kwanza was also closely aligned with the World Economic Forum's New Vision for Agriculture, launched in 2009 to promote "market-based solutions to activate public and private investments" in agriculture. The New Vision was championed by major multinational corporations, including Archer Daniels Midland, BASF, Bunge, Cargill, the Coca-Cola Company, DuPont, General Mills, Kraft Foods, Metro, Monsanto, Nestlé, PepsiCo, SABMiller, Syngenta, Unilever, Walmart, and Yara International.[82]

To put Kilimo Kwanza "in motion," Kikwete subsequently launched the Southern Agriculture Corridor of Tanzania (SAGCOT) at the World Economic Forum on Africa in Dar es Salaam in May 2010.[83] SAGCOT became the first major development initiative in the history of the nation to systematically promote PPPs and plantation agriculture, especially in partnership with major food and agrochemical companies. The geographical scope of SAGCOT would encompass one-third of mainland Tanzania, from the coastal plains to the central valleys to the southern highlands. By 2030, it aimed to bring 350,000 hectares within the corridor under commercial agricultural production; transform ten thousand smallholders into commercial farmers via participation in outgrower schemes tied to plantations, including the proposed EcoEnergy sugar estate in Bagamoyo; create 420,000 new jobs in the agricultural value chain; and generate USD 1.2 billion per annum in revenues from agriculture. The SAGCOT Investment Blueprint was subsequently unveiled in January 2011 at the World Economic Forum in Davos by President Kikwete and the CEO of Unilever. Following SAGCOT's launch, Kikwete in 2013 introduced yet another ambitious initiative, Big Results Now (BRN), to achieve "quick wins" in six national priority areas, including agriculture. By 2015/16, BRN sought to establish twenty-five new commercial farm deals for paddy rice and sugarcane, including the EcoEnergy land deal in Bagamoyo; establish seventy-eight rice irrigation schemes; and build 275 warehouses to improve the marketing of maize.[84]

The momentum for Kilimo Kwanza and SAGCOT grew rapidly in tandem with the rise of other high-profile policy initiatives at the regional and international level. Key examples include Grow Africa, which the World Economic Forum launched in May 2011 in partnership with the African Union to create a "market-based platform" to increase private-sector investments in African agriculture.[85] Kikwete hosted the first Grow Africa meeting in Dar es Salaam in November 2011. In May 2012, he announced Tanzania's cooperation with a USD 8 billion initiative, the New Alliance for Food Security and Nutrition. The G8 (under the Obama presidency), African governments, private corporations, philanthropic foundations, and development agencies initiated the effort "to catalyze responsible private sector investment" in African agriculture, and in so doing lift fifty million people out of poverty by 2022.[86] The New Alliance praised Tanzania as a "showcase for PPP in agricultural growth" in Africa, although there was hardly any evidence to support such a claim.[87]

Both the Kilimo Kwanza and the SAGCOT initiatives bypassed the usual process of national agricultural policy making, which had hitherto involved the leadership of the agricultural ministry together with donor agencies. Instead, Kilimo Kwanza came about as a result of discussions held by domestic agribusiness elites and commercial farming classes and facilitated by the Tanzania National Business Council, a quasi-autonomous state institution housed under the PMO and chaired by the president.[88] As for SAGCOT, it was the Norwegian fertilizer giant Yara International that introduced the idea.[89] The company first pitched the concept of "agricultural growth corridors" at the UN Private Sector Forum in New York in September 2008, and in October 2009 presented it to President Kikwete, who endorsed it. The Norwegian government, a key shareholder of Yara, provided financial support for drafting the SAGCOT concept note that Kikwete released at the World Economic Forum on Africa in May 2010.[90]

The proliferation of these market-oriented agricultural initiatives in the post-2008 period created significant challenges to policy coordination and a disjuncture in the state's vision for rural development. Whereas the Ministry of Agriculture was responsible for the implementation of the ASDP, the PMO became the coordinating agency for Kilimo Kwanza. An independent secretariat managed SAGCOT, while the State House oversaw the BRN. If the ASDP saw private investments as supplementary to public investments, the other initiatives centered heavily on incentivizing corporate actors to supplant the role of the state. Rather than prioritizing the needs of small-scale producers, Kilimo Kwanza, SAGCOT, and BRN catered to the interests of commercial agribusinesses, multinational corporations, and foreign investors, and promoted the incorporation (often euphemized as the "inclusion") of smallholder farmers into the global value chain.[91] It is in this context that the government began representing and holding the EcoEnergy project on a pedestal as a model investment for agricultural modernization.

Assembling Land and Capital

Having successfully reframed its business as a PPP-based development cooperation, EcoEnergy turned to raising start-up capital from the IFIs. In early 2011, the company approached the AfDB's Private Sector Department to request a loan to cover the bulk of its total estimated project cost of USD 542 million.[92] Between 2011 and 2012, as a first step toward fulfilling the bank's due diligence requirements, EcoEnergy hired external consultants to produce necessary documents, including the ESIA and the Resettlement Action Plan, the former of which I discussed earlier and the latter of which I discuss in detail in chapter 3.[93] Beyond complying with these safeguard procedures, one of the most critical conditions for debt financing required by the AfDB was the formalization of land acquisition. As a Dar es Salaam–based AfDB official explained to me in an interview, "No financial agreement would be made without clear proof of [EcoEnergy's] exclusive access to property."[94] The land acquisition process that had more or less stalled since late 2009 had to be resumed quickly.

Exchanging "Land for Equity"

In her annual budget speech to the Parliament in June 2012, the minister for lands, Anna Tibaijuka, introduced publicly for the first time the idea of "land for equity."[95] While promising to establish once and for all the land bank and the land compensation fund, the minister insisted that there had to be a new form of partnership and a benefit-sharing mechanism between the government and investors, both domestic and foreign. She envisioned an arrangement in which the government would lease land to investors, in exchange for equity shares in investments. She called on members of the Parliament to support her idea so that the nation could "use land as capital [*kutumia ardhi kama mtaji*]."[96] Land was to become a financial asset from which not only ground rents but also dividends from land-based investments flowed. In a video interview released in November 2012 and later posted on EcoEnergy's website, Tibaijuka described "land for equity" as a "policy" and a "win-win situation" that would reposition Africa as no longer a continent "to be used," but as a place where Africans themselves could participate in and benefit directly from investment opportunities. It was about time, she said, that the nation did away with past practices of allowing investors to provide ad hoc in-kind benefits like "water wells and classrooms" and instead demanded real financial returns.[97]

Although her policy proposal had yet to be drafted and debated within the Parliament, the minister had already communicated to EcoEnergy that the government was ready to enter into a "land for equity" contract with the company. In a letter to Per Carstedt dated February 23, 2012, Tibaijuka wrote, "I am glad that you have accepted to pioneer our new policy of Land Based Investments [sic]. For an equity share of at least 25% into the venture, the GoT [government of Tanzania] will provide land free of encumbrances to the foreign investor who, in turn, will provide capital management."[98] The letter further specified that in exchange for land, the government would acquire 10 percent of the shares in the project company upon signing of the land lease, and that the shares would increase to 25 percent after eighteen years of project operation. Just as the planning minister had signed an MoU with Sekab in the absence of legal or regulatory provisions, the lands minister, too, put the cart before the horse when it came to the land allocation.

Several government officials I interviewed referred to "land for equity" as "Tibaijuka's baby."[99] On the other hand, a consultant for the United States Agency for International Development (USAID) described it as "Carstedt's baby, which Tibaijuka successfully adopted."[100] Regardless of who originated the idea, a policy on "land for equity" never existed and still does not exist today. Several government officials within and outside the Ministry of Lands expressed their reservations about the way the government negotiated the land deal with EcoEnergy. As the assistant commissioner for lands quoted earlier put it,

> It's unfortunate. It was just a concept. The EcoEnergy project was the first project the government intended to establish as a "model," but unfortunately, we created a model without any policy, institutional, and legal framework to support it. That is the problem. You cannot find any government document about "land for equity." You will not find it. There was a concept note about it, but it did not sail through in the government. There was a lot of confusion on how to implement the model or how we would actually share the profits, etc.[101]

The Ministry's deputy permanent secretary also emphasized that "land for equity" was "just an idea and never a policy."[102] The PMO official I cited throughout this chapter was the most critical. In his view, EcoEnergy had sold the idea of "land for equity" to Tibaijuka to expedite the land acquisition and loan financing from the AfDB. He believed it was inappropriate and risky for the government to pilot not only a fictitious policy but also to do so with a foreign investor who had no proven record of implementing a large-scale agricultural project anywhere in Africa. The company's willingness to give as much as 25 percent was not so much to share the benefits, he argued, but to share the risk that the project might fail:

> EcoEnergy doesn't have a penny! What does it need in order to get money? The title deed! This has been my biggest problem with my colleagues. They didn't fully understand that EcoEnergy needed the title deed to take out loans. They should have thought more carefully of why EcoEnergy was willing to give the government as much as 25 percent. That is because they didn't have anything! They have never started a project in Africa. They could say whatever because they didn't have anything real to show.[103]

Donor agency observers were also cautious in recommending "land for equity" as a benefit sharing mechanism in land-based investments. A 2014 USAID-commissioned report, for instance, emphasized that "land for equity deals are risky," especially in a context "where the minority shareholder has limited financial expertise" and where there is significant "uncertainty over whether the venture will be profitable."[104]

Despite these reservations, this financial sharecropping arrangement was formalized in a certificate of title the government conferred to EcoEnergy on May 8, 2013. Whereas the Land Act and the Village Land Act of 1999 stipulate that land transfers may occur *only after* an agreement has been reached with existing customary land occupants regarding the "type, amount, method, and timing of compensation" and after they have been paid "full, fair, and prompt compensation," the government failed to heed these regulations, arguably because it presumed the land to be bona fide public/general land (see chapters 2 and 3).[105] Whereas foreign investors are typically allocated derivative rights of occupancy in the name of the TIC, EcoEnergy was given a granted right of occupancy in the company's own name, because, according to the assistant commissioner for lands, the government and EcoEnergy were "joint owners of the project."[106] Based on the certificate, EcoEnergy was entitled to exclusive rights to occupy and use 20,373.56 hectares in Bagamoyo District for ninety-nine years, with an annual rent of TZS 50,344,104.[107] In May 2013, the rent amounted to roughly USD 31,000 per year, or USD 1.55 per hectare per year. As I discuss next, however, obtaining the title deed was just one of many hurdles EcoEnergy had to overcome to get the project off the ground.

Risks, Conditions, and Limits of Development Financing

In May 2012, while the negotiations with the AfDB and the Tanzanian government were under way, EcoEnergy turned to Sida to seek additional financial support. Despite its failure to bring the agency on board three years prior, the company applied again for a credit enhancement guarantee, worth USD 94 million. Notwithstanding its prior rationales for rejecting Sekab BT's request, Sida agreed this time, in principle, to approve the guarantee—the agency's largest ever—to fund EcoEnergy's potential cost overruns and/or early revenue shortfalls, contingent on necessary project appraisals.[108] As a Sida representative explained, "We did our preliminary assessment of the request and deemed that the project was worthwhile, though there were some big stones to be turned. We determined that the finance guarantee would be appropriate if development actors like AfDB were involved."[109]

In mid-2013, with the financial agreement with the AfDB still pending but eager to get started on project implementation, EcoEnergy applied for a short-term commercial bridge loan of USD 18 million from the Standard Bank of South Africa (SBSA). In July 2013, the company asked Sida to release the guarantee early so it could use it toward underwriting the SBSA loan. Though project appraisals had yet to be completed, Sida agreed to contribute a stopgap guarantee of USD 16.2 million (90 percent of the SBSA loan).[110] This financing agreement, which went into effect in February 2014, came with several strings attached. EcoEnergy was expected to repay the guarantee once it reached a financial close with AfDB. Should EcoEnergy fail to return the guarantee, Sida would have a legal claim on the company's assets.[111]

In April 2014, the AfDB board of directors approved a loan package of USD 100 million toward establishing the EcoEnergy sugar plantation and factory, and potentially an additional USD 30 million toward the outgrower scheme.[112] These approvals, too, were subject to conditions EcoEnergy and the Tanzanian government together needed to fulfill. The first condition entailed the resolution of all outstanding land disputes within the proposed project site, a topic that I delve into in the remainder of the book, especially in chapters 2 and 6. The second and related condition was the resettlement of local populations according to the bank's operational policies and so-called international best practices, which I discuss in chapter 3. And the final condition was the formalization of a power purchase agreement between EcoEnergy and the state-owned electricity supply company, TANESCO.[113] In late 2014, following a national audit, Sida added three additional conditions to its loan guarantee.[114] The agency required the company to bring onboard another long-term strategic partner who had real technical expertise in developing and operating sugarcane projects. It also asked the company to ensure that the Tanzanian government reformed its sugar policy to protect domestic producers and curb illegal sugar imports. Lastly, it demanded that the company seek confirmation from the AfDB that financial closure would indeed be reached. EcoEnergy had until the end of April 2015 to resolve these issues.

When the deadline arrived, none of the conditions had been met. After having disbursed approximately USD 6.2 million, Sida finally withdrew its financial support from EcoEnergy.[115] SBSA consequently terminated its loan in May 2015 and soon asked the company for repayment.[116] This financial fallout coincided with the release of ActionAid International's global campaign report and online petition titled *Stop EcoEnergy's Land Grab in Bagamoyo, Tanzania*. The report argued that the company had failed to obtain free, prior, and informed consent from local communities, and that its proposed outgrower scheme involved far more risks than benefits.[117] Both the Tanzanian government and EcoEnergy publicly denounced the organization's findings, calling its research "strange," "flawed," and "unethical."[118] According to an ActionAid Tanzania staff member, although the campaign continued at the international level, the controversy and subsequent government pressures made it challenging for the organization to continue its work in Bagamoyo.[119]

Despite this damaging turn of events, EcoEnergy managed to secure a mix of loans and grants worth USD 66.6 million from IFAD in late 2015 to support the outgrower development program. IFAD justified the funding on the basis that the program embodied the national policy priorities on delivering a "private-sector-driven" and "pro-poor inclusive business model" of agricultural development. Mirroring the language the outgrower specialist used earlier, an IFAD report endorsed the EcoEnergy Sugar Project as a "4Ps investment," a public-private-producer-partnership that would serve as a "model for 24 future investments" in commercial farms planned under the BRN.[120] By November 2015, EcoEnergy had also finalized an agreement with a new strategic partner, the Uttam Group, which owned four sugar mills in India. However, Sida's withdrawal, mounting debt, delayed negotiations with the government of Tanzania and AfDB, and growing public opposition—all on the eve of regime change in the country—rendered EcoEnergy's future ever more uncertain.

Whither the Partnership?

When asked to describe Tanzania's relationship with EcoEnergy, the assistant commissioner for lands chuckled and described it as an "unhappy marriage," a metaphor that signaled some degree of violation of legal and moral obligations and expectations the parties had to each other. Noting that the company "had not paid a single shilling in rent," he went on to say, "Here you have a situation where the marriage is not working. Can you ask your partner to buy you a gift? I am telling you, it's an unhappy marriage."[121] From EcoEnergy's perspective, it was not liable for any rent unless the government provided the company land free and clear of encumbrances, that is to say land without people or competing property claims.

The relationship between the Tanzanian state and EcoEnergy became increasingly tenuous throughout 2015. The high-profile politicians who had previously supported the project were losing their political influence. Minister Anna Tibaijuka, for instance, was sacked in December 2014 for receiving a deposit of over USD 1 million in her personal bank account from Tanzanian business elites in the energy sector.[122] And in October 2015, Jakaya Kikwete stepped down as president after ten years in office, following a disappointing overall performance on improving democratic governance and eliminating corruption.[123] A common theme in my interviews with government officials between 2015 and 2016 was that EcoEnergy was too demanding, nonreconciliatory, and naïve about "how things were done" in Tanzania. The PMO official expressed this sentiment best:

> This is my problem with my friends in EcoEnergy. I told them: Look guys, I doubt if you are really listening. You are asking for too much. You know, for instance, you must be very stupid when you ask for sugar policy reform, saying you want the value-added tax to be waived on the sugar that we are going to sell in our own market. No one is going to accept you with all your requests, especially after Kikwete. Our next president [John Pombe Magufuli] will not be from Bagamoyo but from Chato [a district in Geita region in northwestern Tanzania], and he will not be interested in you.[124]

An AfDB official based in Dar es Salaam shared similar views:

> Since 2014, I witnessed a tug of war between EcoEnergy and the Tanzanian government. EcoEnergy was too ambitious in asking for policy changes. They were the ones who wanted the government to review the national sugar policy, particularly importation policy. But you know, reviewing or reforming any kind of policy does not happen overnight. It's a step-by-step process that needs to involve dialogue with different stakeholders, and this can take a very long time. This is primarily why I think the project has dragged on for so long. The investor probably thought this policy [reform] process was going to be carried out quickly, but you know, they actually ended up digging their own grave.[125]

He also indicated that Sida's ambivalent and shifting position on EcoEnergy did not help the situation:

> The Tanzanian government received conflicting signals from Sida over the years. First, many years back, Sida signaled to the government that

> Sekab BT was not a trustworthy investor. The government hesitated to go ahead. Sida signaled to the government again in 2015 that EcoEnergy was no good. But earlier this year [2016], there was another communication from Sida, signaling that the earlier view of Sida was not the "true view of Sida." But imagine, let's say I am a Kenyan investor wanting to invest in Tanzania, but the Kenyan government was informing the Tanzanian government that I was not a trustworthy investor. It's like your father telling someone that you are a bad son. Of course the Tanzanian government hesitated and had to think twice about its decisions.[126]

Sida's assistant director general and head of its Africa Department had indeed written to Tanzania's State House in February 2016 to inform the new president that the agency was now in favor of the EcoEnergy project. With a USD 6.2 million claim on the company and with growing public criticism over misuse of taxpayer money in Sweden, Sida sought to recover its losses, but in ways that arguably created more doubts for the Tanzanian state.[127]

From EcoEnergy's perspective, the project was held back as a direct result of the Tanzanian government's acts and omissions, a subject I will come back to in the conclusion. In an interview in October 2015, Per Carstedt blamed the delay on what he considered the inefficiency, corruption, and lack of political will of the Tanzanian government. I asked whether, based on his experience, he still believed in PPPs as a development model. His response was "This is PPP: Passion, Perseverance, and Plan B." When I asked what he meant, he replied,

> I mean, we've had to do it many times. You agree with the government at one point that this is the rule, this is the map, but then suddenly the map changes, the minister changes. . . . But our compass has always been the same, our value is the same. . . . I mean, the external values of an integrated sugar project like this are many, many times higher than the values you can generate on balance sheets and financial statements. That's why governments around the world are supporting these projects because the long-term external values are enormous. Tanzania [*pause*] hasn't come to this point yet. Okay? And that's why, for us, it has been extremely frustrating. . . . We have done A, B, C on our part, but the government hasn't done their parts D, E, and F. That means we can't go ahead. We have to wait for the next window. Then at the next window, still nothing gets done. . . . Here you have a government that doesn't have capacity and some who are corrupt, they have other agendas, of course things get complicated. . . . The normal Tanzanian solution [to a problem] is that they don't do anything. That's the Tanzanian attitude.

When there's a problem, they tend to stick their heads into the sand, and hope it will solve by itself.[128]

Lay observers might consider Carstedt's views on PPPs and EcoEnergy's permutations over time as demonstrating the necessity of corporate adaptability and flexibility in the face of adverse business and political conditions. Or they might read the EcoEnergy case as a cautionary tale about the administrative challenges of investing in agriculture and farmland in Tanzania. These readings are not necessarily wrong, but they are too simplistic. Focusing on the decade between 2005 and 2015, this chapter has examined the diverse and competing sets of actors, processes, and relationships that simultaneously supported and frustrated the EcoEnergy land deal. Whereas the phrase "land grabbing" tends to conjure up images of all-powerful foreign investors sweeping in and taking land at will, this chapter revealed a much more complex and nuanced story. Contrary to popular belief, investors like EcoEnergy do not always come with capital or technical know-how ready at hand, nor do they find it easy to acquire land in places like Tanzania because land is supposedly "cheap" or because there is "weak" land governance.[129] As I demonstrated, the transformation from Sekab BT to EcoEnergy was made possible, in large part, by the company's ability to hustle and exploit, often in underhanded ways, the shifting political-economic conditions and policy incentives at the national, regional, and international levels and across various sectors, including energy, transport, agriculture, and international development.

What remained constant in the company's metamorphosis (apart from its leadership structure and the focus on sugarcane) was its use of "win-win" narratives as a legitimizing device. These narratives were essentially claims about the future in which there were no losers, only winners: an imagined utopic future where the foreign investor and the state would equally benefit, as would the local communities, even though they would lose their land and may or may not be employed by the project that dispossessed them. These "win-win" narratives and claims of mutually beneficial partnerships worked insofar as they sanitized and concealed the unequal and conflictual power relations between diverse actors differentially positioned within society, the agricultural value chain, and the world economic system at large. On the whole, such narratives only made sense if one assumed the inevitability and superiority of large-scale, monocrop plantation agriculture and outgrower schemes over smaller-scale, diversified, and subsistence-oriented food systems. As I highlighted throughout the chapter, the company's attempts to keep the project afloat did not go unchallenged. They were met with many frictions and Polanyian "double movements" in Tanzania, Sweden, and across the world, in which ordinary citizens, journalists, researchers,

NGOs, and some state officials attempted to protect society and the environment from the unbridled forces of market capitalism.[130]

Of all the different challenges that forced EcoEnergy to reinvent itself, however, one question remained largely untouched: the land question. While EcoEnergy and the Tanzanian government, under the Kikwete regime, found legal and extralegal workarounds to formalize the land transfer, they still had to reconcile and come into worldly encounter with hundreds of existing resource users in Bagamoyo. That is why the contemporary dynamics this chapter chronicled must be supplemented by and situated within the longer history of enclosures in Bagamoyo and the wider Coast Region in Tanzania. In the next chapter, I thus begin my analysis from the very ground on which the land deal stands. The tendency of the Tanzanian state and the foreign investor to gloss over the history of the landscape and the persistence of small-scale food producers, I will argue, explains how and why the land deal remained incomplete.

2
THE MAKING OF A BITTER LANDSCAPE

> It is space, not time, that hides consequences from us.
>
> John Berger, *The Look of Things*

Traversing eastern Bagamoyo along the Swahili coast, one witnesses an array of historical remains. On entering Bagamoyo town, one encounters the Old Fort, which the Omani Sultanate of Zanzibar had built during its dominance over the East African coastal trade in the late nineteenth century.[1] The fort initially housed the governor who oversaw the sultan's business interests, and later the Germans and the British repurposed it as a town prison and police station.[2] Nearby, close to the beach, stands a small obelisk monument in the place of a tree where Africans revolting against German colonial oppression were said to have been hanged between 1888 and 1889.[3] Kitty-corner from the hanging place, the Germans built a new district administrative building in 1897, which both the British colonial and independent governments kept in use for the next century. The Caravanserai, located in the center of town, was a popular nineteenth-century guesthouse that served hundreds of thousands of African porters—men, women, and children, many of them of Nyamwezi origin—who trekked for weeks from the continent's interior to the coastal trade towns like Bagamoyo, Winde, and Saadani, bringing pounds of ivory, gum copal, rubber, and enslaved people to exchange with Arab, Indian, European, and American merchants for cloth, weapons, and copper wire.[4] On the northeastern edge of town near the shoreline sits the Holy Ghost Mission, the first Catholic mission the French Spiritans established in the East African mainland in 1868, against the protests of the local Zaramo, who resisted the enclosure of their land.[5] Until the early twentieth century, the French mission operated a profitable business in Bagamoyo with extensive landholdings, coconut plantations, a cotton ginnery, and a copra drying plant, all of which mobilized African labor in the name of civilization or "the regeneration of the black races."[6]

Historians have traced the name Bagamoyo—a combination of the Swahili words *bwaga* (to throw down) and *moyo* (heart)—to the town's status as a terminus of the central caravan route, a place where porters could finally throw down their loads and relax to their heart's content. Other observers attributed the name's origin to the lament of enslaved people who were forced to leave their hearts and souls behind before being shipped off to Zanzibar and other foreign lands.[7] Either way, urban Bagamoyo was a place of commercial and political significance in the pre- and early colonial period.[8]

Less well apprehended, however, is a relational history of rural life in the district's fertile hinterlands, centered on the Wami-Ruvu river basin. The Wami and Ruvu Rivers originate from the Eastern Arc Mountains in central Tanzania, flowing eastward and forming large estuaries that drain into the Indian Ocean.[9] Lying thirty to forty meters above sea level, the area sandwiched between the downstream sections of the two rivers comprises a complex socioecological assemblage of coastal mudflats, mangrove swamps, alluvial plains, riverine forests, seasonal wetlands, and a mosaic of bushland, grassland, and woodland, interspersed with cultivated fields, human settlements, and wildlife habitats. Along the floodplains of the Wami River, a vibrant agricultural scene unfolds: green maize and rice fields intercropped with fruit trees, legumes, and other vegetables, surrounded by dense reeds, patchy grasses, and shrubby thickets of bushes and small trees. South of the floodplains lies an expanse of coastal plains dissected by smaller rivers and seasonal streams; the gently undulating hills are enlivened with crop fields, grazing land, and homesteads amid thorny acacias and age-old baobab trees. The EcoEnergy Sugar Project would be introduced into this dynamic coastal riverine landscape.

Here, in contrast to urban Bagamoyo, the vestiges of imperialism or what Ann Stoler has called "imperial debris" are much less conspicuous.[10] If found—like the remains of a colonial plantation on the north bank of the Wami River, approximately fifty kilometers from Bagamoyo town (figure 2.1)—they are not memorialized as tourist attractions or nationally protected heritage sites. Yet despite their forsaken appearance, they are not forgotten from the landscape or from people's memory. They are woven into the fabric of everyday life; they are among the many mundane and partially remembered gatherings of the landscape. In their oral histories, local elders would often say "ardhi inaficha sana"—the land hides a lot. By this they did not simply mean to describe how artifacts like the plantation remnants are easy to miss if one is oblivious to their natural surroundings. As I came to understand them, *land* and the act of *hiding* meant something more capacious. The land hides stubborn weeds that lie dormant but alive in

FIGURE 2.1. Remains of a colonial-era plantation in Kisauke on the north side of the Wami River. Photo by the author, August 2014.

the soil; the land morphs into water at certain times of the year to bring bountiful harvests and catches of fish; the land offers spaces of sociality and seclusion during female initiation rites; and the land holds gendered wisdom on how to live and work with nature. Simultaneously, however, the land obscures a morass of ambiguous boundaries, conflicting maps, and overlapping property rights; it conceals the traumas and triumphs villagers faced, and continue to face, as they wrestle with the legacies of old and new enclosures.

This chapter attends to these dialectics of presence and absence, sustenance and suffering, to historicize the landscape in rural Bagamoyo. It examines the shifting patterns of human settlements, land use practices, gendered environmental knowledges, and a series of extralocal political-economic processes that have upended agrarian life throughout the colonial and postcolonial period, to dispel the notion that the land granted to EcoEnergy had been "idle" or "unused." The history I offer is not a backdrop but is at the forefront of precisely what is contested. The fundamental reason the EcoEnergy land deal remained incomplete is that it became inescapably enmeshed in a landscape characterized by long-standing and unresolved conflicts over resource access and control and the ambiguities in land tenure and boundaries these conflicts created. Landscape, as political ecologists and cultural geographers have long argued, is co-constituted with identity, livelihoods, and belonging.[11] It is lived, felt, practiced, and contested.

It is an always-unfolding spatial drama created by material, cultural, and political processes, shaped as they are by competing visions of what the landscape ought to look like and ought to achieve. The main theme of the spatial drama I explore in this chapter is how rural women and men have historically asserted themselves—and in so doing, co-produced the landscape—through *everyday acts of presence*: by staying put against what they were "supposed to do," by using the land on the margins of former plantations, or by reoccupying and reclaiming the land after long periods of dispossession, to ensure their survival and social reproduction.[12] No enclosure could thus be permanent in the face of rural people's desires and struggles to remain on the land from one generation to the next.

Lived Landscape

"Kwanza, choreni mto"—First, draw the river. As elders gathered around for participatory mapping in the village of Matipwili, there was no doubt in anyone's mind about what should appear first on the large sheet of paper I had provided them. Once they drew the Wami River and the Indian Ocean, they marked the railway and the road, and then a drew small box they labeled with the year 1910 on the north side of the river. The box denoted where a Greek sisal plantation used to be, and long before that, a German cotton plantation. After drawing these landmarks, elders began listing the names of all cultivated areas and settlements along the north and south banks of the Wami River. At another mapping workshop in Makaani, south of Matipwili, the elders there followed a similar flow: the river, the ocean, the road, the railway, the 1910 landmark, and fields and homes (see figures 2.2 and 2.3).

EcoEnergy's 2,0373.56-hectare land concession encompasses three administrative village units: Matipwili, Razaba, and Kitame (see figure 0.7). Though Razaba and Kitame are officially subvillages of Makurunge, they assumed a villagelike status after Makurunge became designated as a township in 2010.[13] As I will discuss later, villages like Matipwili and Makurunge are products of postcolonialism, having been established in the 1970s as part of the socialist state's villagization program. Razaba and Kitame were registered more recently, in the 1990s, as Tanzania embarked on neoliberal economic reforms and multiparty elections at the insistence of Western donors. While these places and place names offer useful reference points for examining the co-constitutive nature of the landscape and postcolonial political economy, they are less instructive for understanding the fluidity of social, ecological, and cultural relations through which agrarian life and livelihoods came into being, long before these borders and names were ever invented.

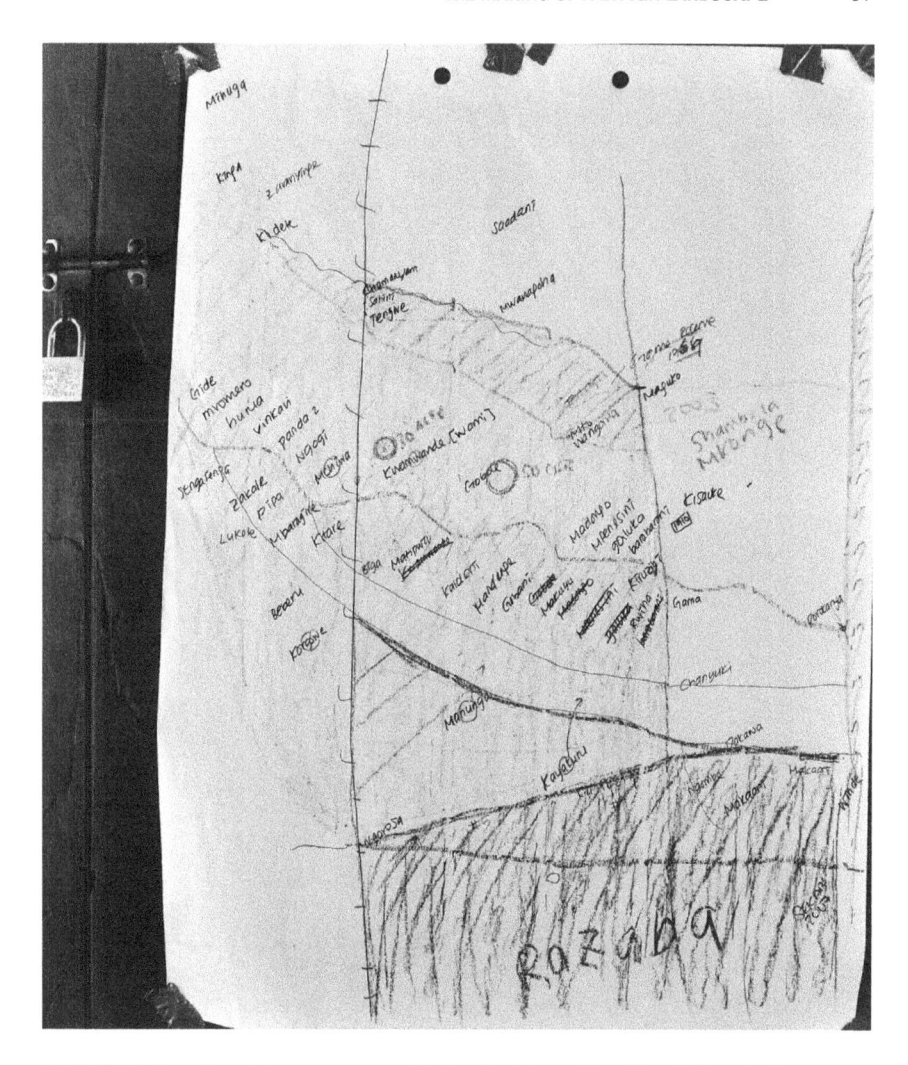

FIGURE 2.2. Hand-drawn map by Matipwili elders. The Wami River is the thicker line that winds across the middle of the page. The Indian Ocean is the colored segment running vertically on the far right. The railway is the vertical line with hashmarks on the left; the road is the parallel vertical line in the center-right. The 1910 box denoting a colonial-era water tower is just north of the river, on the right-hand side of the road. Photo by the author, March 2016.

In oral history interviews and participatory mapping workshops, elders used the word *Wami* to denote not just the river but the broader fluvial landscape, inclusive of the valleys, floodplains, forests, and other dwelling places (figure 2.4). Wami was "sehemu yetu ya zamani"—our old place. Sometimes, the elders used Wami interchangeably with *Udoe*, the Doe territory, "tribal division" or "native authority" as

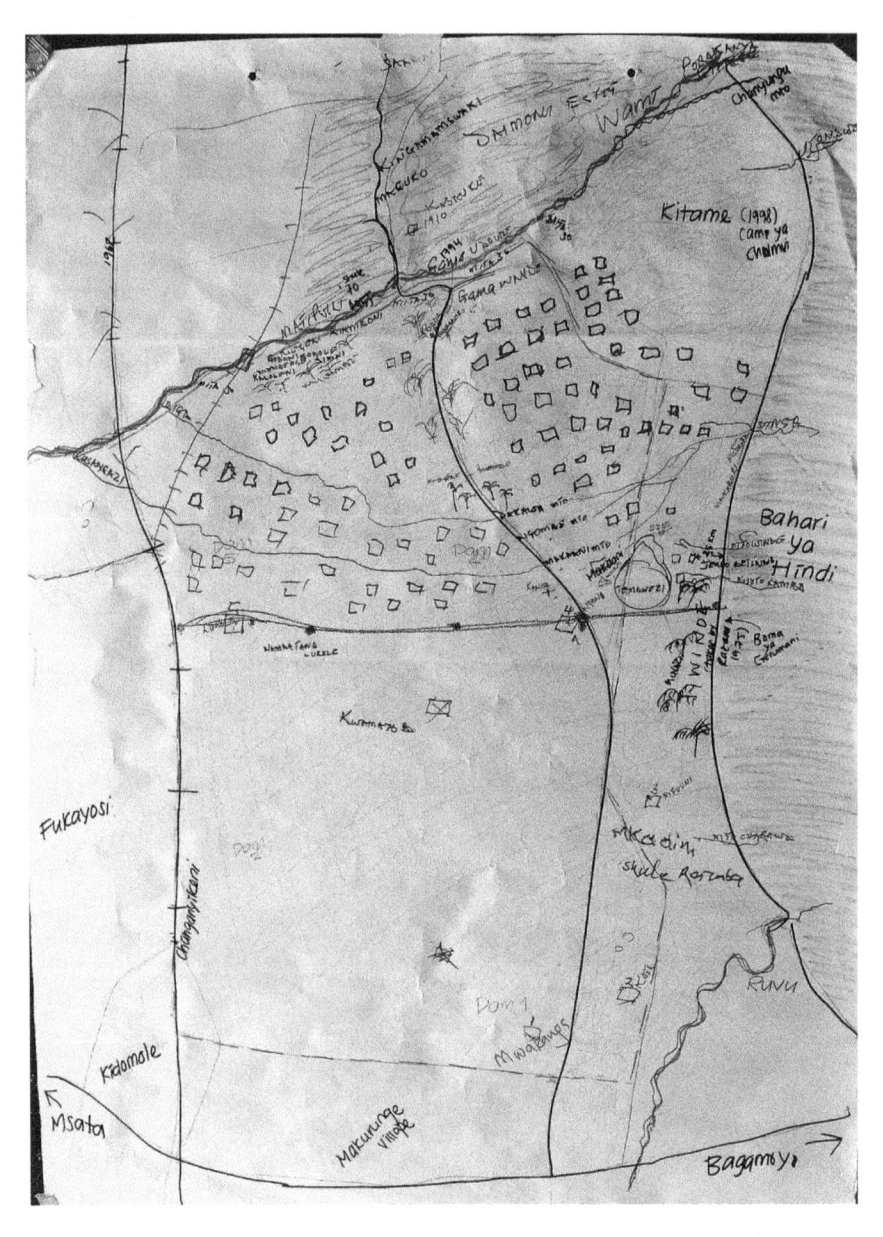

FIGURE 2.3. Hand-drawn map by Makaani elders. The Wami River is the dark line that runs diagonally across the top of the page. The Indian Ocean is the shaded area on the far right. The railway is the vertical line with hashmarks on the left; the road is the nearly parallel vertical line in the center-right that curves slightly northwest. The 1910 landmark is in the center at the top of the page, just north of the river, on the right-hand side of the road. Photo by the author, February 2016.

FIGURE 2.4. The Wami River and the agricultural landscape in Matipwili. Photo by the author, August 2014.

it was known during the colonial period, although the boundaries among coastal ethnic groups, including the Doe, Zigua, Kwere, Zaramo, Luguru, Kutu, and even the Nyamwezi in precolonial times, were fluid, and their socioeconomic relations closely intertwined.[14] According to oral histories, the Doe also intermarried with the Makua, another coastal ethnic group from southern Tanzania (Mtwara region) and northern Mozambique. It is said that three Makua men, Funditambuu, Kimalaunga, and Chamsulaka, arrived in Bagamoyo during the reign of the first sultan of Zanzibar, Majid bin Said (1856–1870), to hunt elephants and participate in the growing ivory trade.[15] Funditambuu was an exceptionally skilled hunter and is said to have taught the local Doe and Zigua men how to hunt. Of the three Makua men, only Funditambuu had come with his wife; but because she was barren, he married a daughter of a local Doe chief. The couple gave birth to a son, to whom many farmers in Wami traced their ancestry. The early Makua hunters initially occupied the area of Makaani (a name derived from *kaaeni*, meaning "stay"), but their descendants settled closer to the river, an area better suited for agriculture and fishing.[16]

Weeds, Floods, and Becoming "People of the Valley"

Siblings and relatives in Wami cultivate plots adjacent to one another or on opposite riverbanks. This spatial pattern of farm organization emerged organically over time as people's livelihoods co-evolved with the life of the river, whose cycles of flooding, erosion, deposition, and meander formation have produced the unique morphology of the floodplains. Through generations of experimentation, floodplain farmers in Wami have developed a farming system adapted to the complexities of the riparian ecosystem. Two farming practices stand out, each shaped by people's material relationships with weeds and floods.

First, Wami farmers engage in shifting cultivation (*kilimo cha kuhamahama*), not necessarily to allow the land to lie fallow to restore soil fertility, but to cope with an invasive weed species that thrives in the wetlands. Known locally as *ndago*—nut grass, or yellow nutsedge (*Cyperus esculentus* L.)—the species is considered agronomically to be one of the worst weeds in the world, particularly for crops like maize, sugarcane, and cotton. It grows up to ninety centimeters tall (almost three feet) and produces an extensive system of fibrous underground roots, which can survive extreme conditions and from which the species reproduces itself. Once established, ndago is difficult to control unless the entire network of basal bulbs, tubers, and rhizomes is completely removed from the soil. Though it is difficult to trace the weed's origin in the area, several elders saw it as an ecological legacy of the plantation agriculture European settlers practiced during the colonial period. Some farmers, typically men, engage in the labor-intensive task of ndago removal if and when there is sufficient labor power. In most cases, however, farmers tend to shift to another place in the floodplains and return later, with the understanding that ndago tubers, including those that lie dormant in the soil, typically have a life span of up to three years, though they may be viable for ten years or more.[17] As Juma, a male farmer born in Wami in 1968, explained,

> Here, the more you farm, the more ndago grows. The land hides many things. You have to experience and learn from the habits of the soil. And when it is time, you must shift. It is something that we became used to, crossing the river back and forth. For a few years we would farm here on this side of the river; then we would move to the opposite site, and then back here again. We would plant trees like mango, wild plum, banana, guava, and coconut to mark our place. If my family shifts to another place, and if someone else comes to my land, they can use it for the time being, but they must respect my trees. That way I can still enjoy the fruits of my land.[18]

His older sister Mwajuma, who was born in 1955, made similar observations: "Almost everyone in Wami, if they have lived here for a long time, they must have farmed on both sides of the river at one point or another. We would farm in one place for five to six years, then shift to another field when ndago starts growing.

The more you farm, the more these grasses will germinate. You must let them live and be prepared to shift."[19]

In the Wami floodplains, the practice of shifting cultivation has shaped, and has been shaped by, customary norms of land tenure where rights of access are conferred to individuals based on occupancy and use. As Juma highlighted above, however, rights of ownership are based on the planting and presence of trees. Mwajuma described early tree-planting: "Our ancestors dwelled first in Makaani, then they moved back and forth across the river, and along the way they planted trees. In Mfenesini they planted a jackfruit tree [*mfenesi*], in Mkoroshoni they planted a cashew tree [*mkorosho*]. Every time they shifted, it was like a celebration; as they moved carrying their luggage, they would sing and play the drums [*ngoma*]."[20]

Though all coastal ethnic groups in Bagamoyo except the Doe have traditionally followed matrilineal inheritance practices, tenure arrangements today tend to be guided by pragmatism about who needs the land, more so than by gender.[21] In Razaba subvillage, in contrast, patrilineal ethnic groups are more present because of histories of migration, which I discuss later, and fathers tend to perceive sons as primary beneficiaries, though often their wives, many of whom are of coastal origin, disagree.[22]

The second practice that defines farming in Wami is flood-based or flood-recession agriculture (*kilimo cha mafuriko*). Understanding the typical Bagamoyo farming system helps us see the distinctiveness and significance of this practice. Following a bimodal rainfall pattern, farmers in the district begin their annual agricultural cycle with the onset of short rains (*vuli*) around mid-October or early November. Maize is the most widely grown cereal in the district, followed by paddy rice. Cereals are intercropped with legumes, such as cowpeas, pigeon peas, and beans. Depending on the soil condition, people also grow tubers like cassava and sweet potatoes, and a wide range of tree crops, including, but not limited to, banana, mango, orange, cashew, coconut, papaya, guava, jackfruit, and custard apple. Women also maintain vegetable gardens where they grow tomato, okra, African eggplant, onion, cabbage, pumpkin, pumpkin leaves, and amaranth leaves. Division of labor varies by households, but in married households men typically clear the land, and women sow the seeds; the work of weeding and harvesting is often shared, though it is not uncommon for women to spend more time on weeding than their husbands do. The first harvests are ready by December and January. Men often perform post-harvest threshing, while women and children assume the responsibilities of home-based processing, including drying, winnowing, and bagging. Around March, farmers prepare the fields again for the long rainy season (*masika*), which lasts until May or early June. Harvesting and post-harvesting work is completed by July and August. Farmers leave crop residues in the fields throughout the dry season (*kiangazi*) to help regenerate the soil until the next planting season begins again in October. During kiangazi, famers engage in various income-generating activities. In Razaba, for instance, women sell home brews, fried fish, and woven mats and baskets, while men engage in various forms of seasonal wage work, such as brick making or working in nearby salt mines or on larger farms.

In Wami, however, the so-called dry season looks very different and also goes by a different name, *kitopeni*: the muddy season, or a spacetime of fertility and growth (*tope*, meaning mud).[23] Though there is no rain, farmers are able to plant again, thanks to the high levels of moisture and nutrients the soil retains after the inundation and recession of the floodplains during masika. The annual floods bring thick alluvial sediments from the upper catchment of the Wami River; the silt and organic matter deposited in the soil acts as a natural fertilizer, allowing people to continue growing food in what would otherwise be a dry season. In kitopeni, women and men alike take advantage of seasonally flooded streams, ponds, and ditches to catch fish, which women gut and prepare for household consumption or for sale. People also introduce paddy rice to their crop mix, in addition to maize. Rice is commonly perceived as a "women's crop," and women tend to exercise more control over its production than men do. As female elders explained, this connection between women and rice emerged during the colonial period when men were forced to take on seasonal wage work in nearby plantations or elsewhere in Tanga, Dar es Salaam, and Zanzibar to meet the demands of taxation.[24] While male labor migration doubled women's workload, it also placed household food production at the center of women's power; and women found ways to save time and labor by occasionally pooling labor with female relatives.

For a visual comparison of kitopeni and kiangazi, consider figures 2.5 and 2.6. Halima and Neema, female farmers in Matipwili and Razaba, took the photographs around the same time of the year, in early August. Given the variations in their place-specific ecologies, Wami farmers like Halima can grow a more diverse mix of crops year-round than their fellow villagers can in Razaba, although Razaba farmers have the advantage of growing other crops that are better suited for dry and sandy soils, such as cassava and sweet potatoes. It was rare for people to go hungry in Wami, Halima said, save for in exceptional circumstances like cholera outbreaks or flash floods caused by the Indian Ocean dipole, which tends to occur every twenty years or so—though with climate change, such events could become more frequent.[25] When I showed Neema's photograph to Halima and other farmers in Wami, they expressed what a boon it was for them to dwell "on water" and to be able to grow food year-round as a result.[26] As Halima put it, "We are coastal people. We live here in the river basin where there is a lot of

FIGURE 2.5. Halima's farm in Matipwili in *kitopeni* (muddy area/season). Maize is intercropped with banana, mango, and wild plum trees. Photo by Halima, September 2016.

FIGURE 2.6. Neema's farm in Razaba during *kiangazi* (dry season). Dried maize stalks are waiting to be cleared for the next planting season. Photo by Neema, September 2016.

water. Water is land, and land is water. Water brings fertility; we can plant almost anything on our land."

Mwajuma, quoted earlier, photographed figures 2.7 and 2.8. She took the first photograph of her field in early May after the recession of the annual floods, and the second in mid-August. As we compared the images, she described kitopeni:

> In Wami we don't have kiangazi. I took this photo [figure 2.7] after the floods. The soil looks cracked, but in fact it is still very moist, you can feel it when you touch it. In the river valleys, floods happen every year, we are used to it. We are people of the valley. These floods are not big floods, although we have had bigger ones in the past, like the one in 1976 and then in 1995 which stayed for three months. As long as the soil is moist and the sun continues to shine, we farm. The food we harvest in kitopeni can last us five months. Maize, rice, pumpkins, fruits, vegetables.... Even if we don't grow more food during this time, we can go fishing. The only food we really need to buy is [wheat] flour, sugar, and salt. Losing this land will cost us a lot.[27]

As Halima, Mwajuma, and many others in Wami have highlighted, land and water were not isolated biophysical elements of the landscape, nor did they consider weeds and floods as disturbances, contrary to how EcoEnergy project documents described them.[28] Land, water, weeds, and floods were not only critical to the ecological functioning of the river and the riverine landscape, but they also

FIGURE 2.7. Mwajuma's farm in Matipwili after the recession of the annual flood. Photo by Mwajuma, April 2016.

FIGURE 2.8. Mwajuma's farm in kitopeni. Maize is intercropped with okra and pumpkin. Photo by Mwajuma, August 2016.

shaped people's livelihoods, identities, and cultural traditions as "coastal people" or "the people of the valley." And as the discussion below will further illuminate, these material and symbolic relationships carried important gendered meanings.

Trees, Forests, and Becoming "Coastal Women"

Forests and woodlands in the coastal plains are home to numerous indigenous tree species on which people depend for sustenance and cultural reproduction. Several elders described these trees as "miti ya mungu"—God's trees.[29] No one knew who planted them, but they offer a myriad of affordances year after year. Though several men, including male healers, were knowledgeable about these common resources, many more women could describe their uses and benefits in greater detail by virtue of their everyday material and cultural relationships with them. Women forage edible fruits; cut fuelwood; harvest leaf fibers; collect medicinal leaves, roots, barks, stems, bulbs, flowers, and seeds; gather resources for cultural rituals; and pass on these knowledges to the next generation. These tasks are all feminized work, but they serve as one of the few affirmations of coastal women's power, knowledge, and status.

Mkole illustrates this. The meaning of mkole is twofold. First, it refers to a multistemmed shrub or what people referred to as a small tree (*mti mdogo*) that produces flexible branches, shiny green oblong leaves, and golden yellow flowers that bear small, round, edible fruits. The tree is found across Eastern and Southern Africa, particularly along the sandy riverbanks and calcareous soils along the coast.[30] Second, mkole refers to a set of female initiation rites practiced by traditionally matrilineal coastal ethnic groups, such as the Zaramo, Kwere, Zigua, and Luguru.[31] Though actual practices vary across clans and villages, the ritual is tied to the mkole tree, which represents fertility and matriliny, symbolisms said to have derived from both the materiality of the tree and the habits of matriarchal elephants. Female elephants are said to hide under an arch made of mkole branches during menstruation and gestation. The pliancy of branches symbolizes female reproductive organs, while the white sticky sap signifies reproductive fluids, including breast milk, cervical mucus, and semen.[32]

For coastal girls, the onset of menstruation sets in motion a series of ritual performances and lessons on female sexuality and morality. In preparation for this period of seclusion and instruction, a *kungwi* (female initiation instructor; pl. *makungwi*) collects a mkole branch and ties it inside the house, usually by the threshold.[33] Mwanahamisi, a Kwere woman who documented the coming-out ceremony of her niece in figures 2.9–2.11, said, "You can use other tree branches, but it won't have the same effect. Mkole can survive in harsh conditions like in the sandy fringes of the river, and so it means it will protect the *mwali* (a girl / female initiate) "from any forces that could make her infertile."[34]

While seclusion is said to have lasted several months in the past, it now lasts a few weeks or up to a month. With the prevalence of formal education, seclusion is also not timed precisely with the onset of menstruation, but typically occurs at the end of the short rains and during school breaks in December or at the completion of the girl's schooling, as a way of publicly announcing her preparedness for marriage. During this time, the makungwi, with the assistance of mothers and aunts, prepare a series of instructions (*mkoleni*; literally "at the mkole tree") for girls who have reached puberty that year. The teachings are highly secretive and often take place in riverine forests (also referred to as *mkoleni*). As a Zaramo woman in her early twenties recalled, "We had to find a hidden place with a mkole tree. The makungwi and female elders gave us many teachings there, everything we need to know about being a woman."[35] Through songs and dance, girls learn about the physiology of the male and female body; sex, sexuality, and desire; different methods of sexual intercourse; and other matters related to marriage, pregnancy, childbirth, breastfeeding, and child rearing. They are also introduced to a variety of medicinal plants that can help with menstrual pain, heavy bleeding, vaginal discomfort, sexually transmitted diseases, fertility enhancement and prevention, postpartum care, and breast milk stimulation.

FIGURE 2.9. *Mkole* (female initiation ceremony). The initiand's (*mwali*) physical crossing of the threshold of a door, her dress and hairstyle, and her subsequent unveiling symbolize her entrance into womanhood. Photo by Mwanahamisi, December 2015.

FIGURE 2.10. *Maulidi* (Islamic religious celebration). Photo by Mwanahamisi, December 2015.

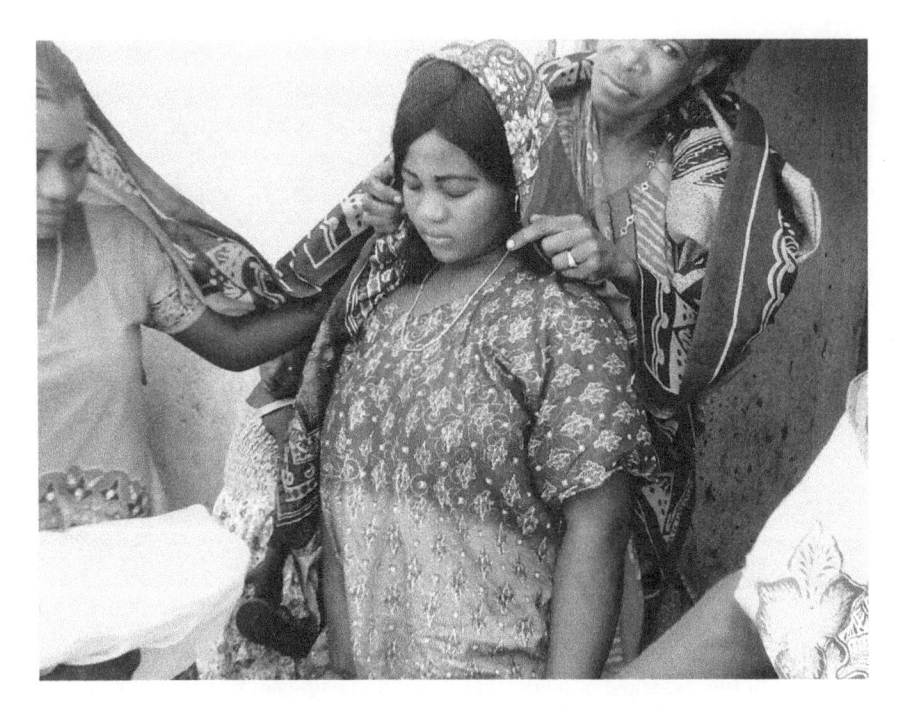

FIGURE 2.11. Unveiling of the mwali. Photo by Mwanahamisi, December 2015.

This private ritual is followed by an *ngoma*, a public celebration involving drum and dance that is open to everyone in the community. It is often accompanied by *maulidi*, an Islamic religious celebration involving tambourine music and the recitation of devotional poetry and verses from the Koran.[36] Afterward, the mwali is escorted outside the house of seclusion under the protection of a *kanga* (colorful cotton wax print fabric) held high up by her older sisters and female adult cousins. The passing of the threshold, the particulars of her hairstyle and dress, and her subsequent unveiling by the kungwi constitute a symbolic announcement of her entrance into womanhood.

In describing the ritual experience, Mwanahamisi emphasized how mkole was indispensable to the construction of coastal womanhood: "If a woman hasn't experienced mkole, then she is not considered proper. We say she must be missing something. Say she gives birth to a daughter later. Even if that daughter is given all the right teachings, she still may not be considered proper because of her mother." Mkole—both as a material resource and cultural practice—embodied the resilience of coastal women in preserving the traditional ecological knowledge that affirmed their identity and status, while sustaining intergenerational female solidarity and sociality.

To summarize, these rich agricultural and cultural practices are, and have been, co-constituted with the specificity of local ecologies. This intertwined socioecological system has given life and meaning to the rural landscape in Bagamoyo and to the people inhabiting and co-creating it. As I signaled throughout the section and as I elaborate below, this landscape was also a direct material result of historical contestations over resource access and control: the struggles of rural people to maintain their relations with the resource base in the face of extralocal pressures on their livelihoods and lifeways from colonial times to the present.

Contested Landscape

As elders drew the 1910 landmark in Kisauke on the north side of the river, a debate erupted over what that structure was, which Europeans had owned it, and whether villagers could still access land in the area. As the elders went on to draw the various tributaries of the Wami River as well as swamps and dams, another debate unfolded around boundaries—boundaries not only between existing villages and the planned EcoEnergy sugar plantation, but also between the villages and a former state cattle ranch and a coastal protected area: RAZABA (Ranchi ya Zanzibar Bagamoyo) and Saadani National Park, both established in the postcolonial era (see figure 2.12).[37] Though the mapping workshops were held in two separate places, the debates that emerged from each were uncannily similar. To facilitate the discussion, I shared with the elders the various maps I had gathered over the course of my research, dated between 1900 and 2014: colonial-era maps of eastern Bagamoyo; maps of protected areas; a map of the former ranch; and maps and irrigation plans of the EcoEnergy Sugar Project. The average age of the forty-five elders who participated in the two workshops was sixty; most of them had spent their childhood during the late colonial period. They had literally and figuratively "grown up" with Tanzania.

Though the maps I shared (and the various visions that were projected onto those maps) directly impinged on people's lives and livelihoods, none of the elders recalled having seen them before. "Ramani zinababaisha!"—Maps are confusing!—shouted one village elder. Others quickly chimed in. "Sisi [ni] wanyonge!"—We are the oppressed! "Tunaumia sana!"—We are hurting a lot! "Tumedanganywa!"—We have been deceived! As one elder explained in more depth in an interview following the mapping workshop in Matipwili,

> The government and foreigners come to us, the citizens, thinking that we are ignorant. They say, "Look, we have the maps! From here to here, it is RAZABA, and the RAZABA land was given to Sekab, and Sekab

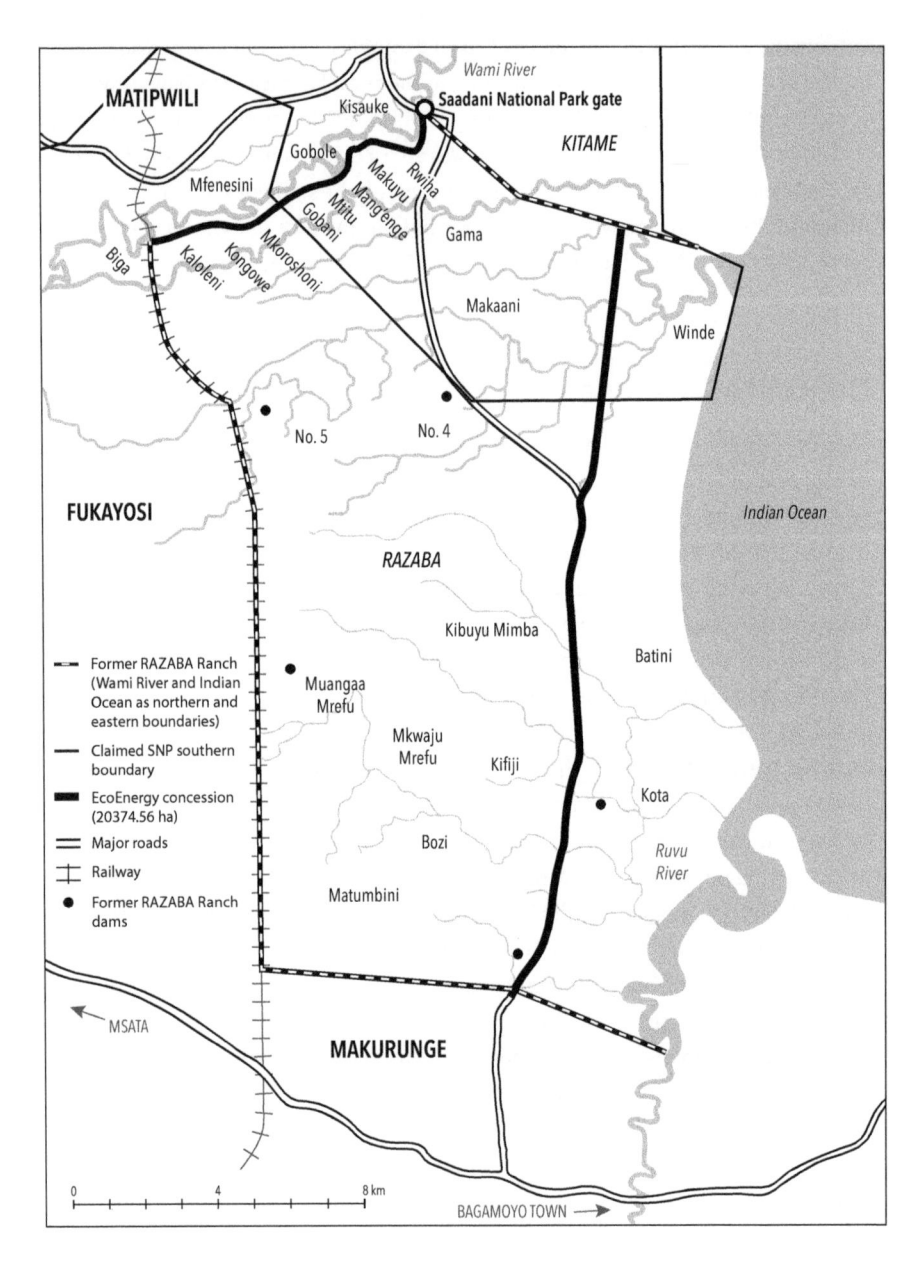

FIGURE 2.12. Overlapping land claims between Matipwili, EcoEnergy, the former RAZABA ranch, and the Saadani National Park as of 2016. Map by the author, based on various official and participatory sketch maps collected during fieldwork.

was given to EcoEnergy. And then there is this national park. We did the surveys and took pictures from the airplane, no one was living here then! It was just an abandoned forest. You all moved here just recently!" When you go to the district government and ask questions, they will tell you: *"We are the government. We have the power."* The government and foreigners like to do their businesses in the air and on their computers. They lie to us. You would expect a lot more from educated people; they are too arrogant.[38]

The most common refrain the elders repeated during the mapping exercise and oral history interviews was "serikali ni serikali"—government is government. This was not so much a statement of resignation or acceptance of state-perpetrated injustice but rather reflected complex and deep-seated feelings of regret, betrayal, bitterness, and suffering. Beneath the vibrancy of the lived landscape were thick, multilayered histories of enclosure, dispossession, and struggles for repossession.

Colonial Enclosures (1888–1960)

The colonial-era water tower, embossed with the numbers "1910," stood on a roadside, surrounded by overgrown trees and bushes, atop the bluffs of the Wami River. It was an important landmark and historical point of reference for many elders. Just barely visible to the left of the year mark are three raised letters: "LBS," an abbreviation for Leipziger Baumwollspinnerei, or the Leipzig Cotton Mill (see figures 2.1 and 2.13).

PLANTATION COTTON

LBS was founded in Leipzig, Germany, in 1884, and in 1907 the company established what was then one of the largest cotton plantations in German East Africa, on approximately thirty thousand hectares between Wami and Saadani.[39] The plantation was established in the wake of the Maji Maji uprising (1905–1907), in which German forces violently crushed hundreds of thousands of Africans revolting against the imposition of the communal cotton scheme and the consequent famines.[40] Between 1902 and 1905, the German colonial government required each village in targeted districts to grow approximately five hectares of cash crops per year. Though initially only male adults had to work in the communal fields, often with little or no wages, by 1904 the increased demand for cotton meant that everyone, regardless of gender, status, or age, was forced to do so. This scheme inevitably led to a crisis of social reproduction for peasant households. Men evaded forced labor by migrating away for other wage work,

FIGURE 2.13. "The Wami River at Kissanke [sic]," German East Africa, c. 1906–1914. Though the larger building atop the bluff no longer exists, the water tank still remains in the landscape.

Source: 105-DOA0596 / Walther Dobbertin / CC-BY-SA 3.0 DE / German Federal Archives.

while women stayed behind and retrenched household food production in ways that saved labor, such as decreasing farm sizes, replacing maize and rice with less labor-intensive and more hardy crops like cassava, and organizing collective work parties.[41]

The end of the Maji Maji uprising in 1907 coincided with a cotton crisis in Germany that forced many industrialists out of business. As competition with American cotton producers intensified, textile industrialists in Germany, including LBS, backed by German banks, pushed to transform African colonies into an agrarian frontier. At the time, LBS optimistically predicted that it could produce all its annual demand for cotton, some thirty thousand bales, on the plantation in Bagamoyo's hinterlands. To that end, LBS introduced European bookkeepers, machinists, water engineers, irrigation, steam tractors, and other expensive capital inputs.[42] Recalling stories that earlier generations had told them, local elders described how existing fields and the commons between Gama and Kisauke, and much of the area north of the Wami River toward Saadani, were "taken by white people."[43] Female elders referred in particular to the enclosure of a stream near Kisauke, called Mto wa Ngoma (literally, the river of drum and dance), where cultural rituals like mkole once took place.

Besides land and capital inputs, LBS's plantation required a regular supply of labor. Yet, fearing another rebellion, German colonial officials averted forced labor policies, which gave African peasants the leverage to refuse work on the plantations.[44] To overcome this problem, LBS turned to recruiting cheap migrant laborers. According to local elders, the name Kisauke originates from this time, when the influx of male migrant workers caused competition for the relatively small numbers of women.[45] As one elder described, "There were many Makonde men [from Mtwara region in the southeast] and men from Burundi who came to work here. They were many men but not enough women, so the fighting began. People began calling the plantation area 'Kisa Mke Kisa Mke' [*kisa* meaning because of, and *mke* meaning woman or wife]. The Europeans failed to pronounce this, so eventually it became Kisauke."[46]

The use of migrant labor, however, seemed to have had a negligible impact on LBS. By the end of 1910, the company had cultivated only approximately sixteen hundred out of over thirty thousand hectares, and by the end of 1911 a mere eighty-three hectares remained in cultivation.[47] The plantation proved unable to deal with floods, crop diseases, soil erosion, and the loss of topsoil caused by heavy steam plows.[48] A common theme that emerged in oral histories was how colonial actors—and later postcolonial ones—had failed to understand the "habits of the Wami," including how the floods, the floodplains, and the wider riparian landscape formed the basis of subsistence and social reproduction.[49] Reflecting on LBS's botched experiment in Bagamoyo, historian John Iliffe remarked, "Plantation cotton showed that capital and technology alone made little impression on Tanganyika's environment."[50] Both labor and nature defied easy commodification.

PLANTATION SISAL

Though the years under British colonial rule are generally described as the golden days of sisal in Tanzania, the fate of the Greek sisal plantation (the Wami Sisal Estate) was not much dissimilar to that of LBS's experiment with plantation cotton. According to archival records, Greek men by the name of Drossopolous acquired the land from the British colonial government in 1926, following Germany's defeat in World War I. Through the Enemy Property (Disposal) Proclamation of 1920, the British sold all land that had been alienated by the Germans, including the area granted to LBS, to European settlers and a few Indians.[51] Once Tanganyika became a British mandate territory under the League of Nations, the British enacted the Land Ordinance of 1923, declaring all land of Tanganyika Territory, whether occupied or unoccupied, as "public land." All public lands became controlled by and subject to the disposition of the governor for the "common benefit, direct or indirect, of the natives."[52] Whereas the Germans during their colonial period had declared all land as crown land under the sovereign control of the German Empire, the British used the 1923 ordinance to legalize, or provide an aura of legality to, land alienation under the pretense of the "common good" of the Africans. Though the ordinance was later amended to recognize the customary titles or "deemed right of occupancy" of African peoples, their land rights continued to be considered inferior to statutory rights or as "tantamount to rights of a squatter," the enduring legacies of which I discuss further below and in the next chapter.[53]

Similar to LBS, the Wami Sisal Estate faced persistent challenges of labor recruitment. As the Bagamoyo district officer wrote in his annual report in 1925, "The local natives have a strong objection to working for anybody else provided they can grow sufficient crops for their own consumption and tax."[54] Elders, too, said local men rarely worked in the plantations unless their proceeds from cash crops were insufficient to pay colonial taxes. As children growing up in the late colonial period, elders remembered watching the *majumbe* (local leaders/headmen; sing. *jumbe*) collect taxes in wooden chests and then carry them atop their heads to deliver to the *liwali* (governor) in Bagamoyo town.[55] According to the Hut and Poll Tax Ordinance of 1922, all able-bodied men over sixteen years of age were required to pay taxes to the colonial government in cash. Married men were forced to pay taxes based on the number of huts or wives they "owned," which entrenched patriarchal structures and ideologies in the most intimate spaces of the home and the family. Tax revenues, however, were difficult to generate, as many resisted the tyranny of taxation. So weak was tax collection in Bagamoyo in 1926 that the governor ordered the "reduction of chiefs' emoluments by half until better results were obtained."[56] As one male elder recalled, "When the *jumbe* came to collect taxes, all the men went hiding in the swamps and forests. When caught, they were either imprisoned or forced to work on improving roads for the colonial government."[57]

So bad were the conditions of production that by 1931 the Wami Sisal Estate exhibited "a depressing appearance of neglect and decay."[58] Though the colonial government entertained the idea of increasing wages to assuage the labor problem and boost production, the district officer remained skeptical. "It is doubtful," he wrote in 1936, that "even higher wages would attract the local native, who prefers to work for himself."[59] In the elders' memory, the plantation fared somewhat better in the 1950s with improved sisal prices, but the estate eventually shut down after Tanganyika's independence. "The Greeks were forced to leave the country after the British went away," recalled one elder.[60]

On the heels of independence and in the absence of formal land redistribution, people began to resettle in Kisauke based on prior customary claims established through land clearance, occupation, and use. As one male elder explained,

"After the sisal estate shut down, people came back to Kisauke and planted cashew trees and mango trees. It became a vibrant place with markets. The 1910 water tank was at the center of all this. People who had fields in the south bank of the Wami built houses there in Kisauke. It was a nice hamlet. A school was built in between Gama and Kisauke to educate the children."[61] A female elder recounted a similar story. After the sisal estate closed, she said, "people from Winde, Tengwe, Saadani came to Kisauke and Gama to buy maize, rice, and vegetables. Some people called Kisauke 'Sokoni' [the market]."[62] Elders remembered the Kisauke and Wami of the early 1960s with fondness and nostalgia. They had reclaimed what had previously been stolen from them; they made Wami their place again by planting trees, building homes, opening markets, and sending their children to school. Now, this vibrant hamlet could only be imagined in people's memories. The abandoned colonial water tank, neglected fruit trees, overgrown vegetation, and wildlife sightings in Kisauke collectively provided haunting impressions of dispossession, repossession, and dispossession redux. As I discuss next, people's desire to rebuild their lives on the land free from colonial oppression would directly compete with the postcolonial state's desire to build the nation through compulsory and arbitrary land acquisitions that ultimately replicated the violence of colonial enclosures.

Postcolonial Enclosures (1961–2010s)

When asked to name one historical event that has had the greatest impact on their lives since the end of colonialism, elders frequently cited Operesheni Pwani (Operation Coast), the code name under which compulsory villagization was enforced in the Coast Region in 1973. As noted in the introduction of this book, villagization was the crux of Julius Nyerere's African socialism: *ujamaa*.

UJAMAA VILLAGIZATION

Nyerere's political program, as outlined in the Arusha Declaration of 1967, included the nationalization of the principal means of production: land. This executive dominance or "trusteeship" over land was a carryover from the colonial legal framework that deemed all land in the sovereign territory public land.[63] Though ujamaa bore similarities to other socialisms like Maoism, especially the primacy placed on agriculture and the role of the peasantry, it was distinctive in that it called for a return to "traditional African socialism" founded on the principles of familyhood and communalism.[64] Nyerere believed that rural life and agricultural production should be spatially organized into communal villages, and this vision eventually spawned the proposal for a nationwide villagization campaign, Operesheni Vijiji (Operation Villages). Though villagization was voluntary between 1967 and 1972, it was made compulsory after 1973, eventually becoming, by most accounts, the largest forced resettlement scheme in postcolonial Africa.[65] By the end of 1976, over thirteen million people were resettled to over seventy-five hundred state-designed villages across the nation.[66]

In the late 1960s, elders in Wami recalled receiving increased pressure from district government authorities to move from the low-lying floodplains on the south bank of the Wami River to higher ground on the north side near the Wami railway station, which opened in 1963. As one elder recounted, "Government workers told us that floods were dangerous and made it difficult for cars to pass, so we shouldn't be living in the valley." People were encouraged to build new houses in concentrated settlements near the Wami train depot so that they could benefit from targeted delivery of agricultural extension and social services. While some people moved voluntarily and built permanent homes on the north side of the river, my interviews with elders suggest that this was due less to their commitment to ujamaa and more to the practical need to prepare for periodic flash floods. Though people were accustomed to annual floods, elders' memories of the landscape were also punctuated by flash flood events that occurred roughly around 1955, 1963, 1976, and 1995, some of which had claimed lives. Over time, it became habitual for floodplain farmers to move and take shelter on higher ground during periods of heavy rains.

When villagization became mandatory, everyone dwelling on the south side of the river up to Gama was forced to move to the north and join the newly formed Matipwili village. Alternatively, they could move south and join Makurunge village closer to Bagamoyo town. Across the nation, the state enforced villagization as a military exercise with considerable use of force, mobilizing thousands of civil servants, police, armed forces, civilian militia, and politicized male youth to move people.[67] In Wami, those who refused to move faced harassment and intimidation by armed forces and risked having their houses burned down or their properties destroyed. "Serikali ni serikali," elders repeated. "It was something Nyerere used to say. 'Serikali ni serikali.' The government has a job to do, and the job can't always make everyone happy," said one male elder.[68] Another elder explained, "Operesheni Pwani was no different from colonialism. There were no proper procedures from the beginning on how to displace people, so what do you expect anything good to come out in the future? All you get is chaos. Serikali ni serikali. That is what Tanzania learned from the Germans and the British."[69]

Once resettled in the new villages, people were required to engage in communal farming using pumped water for irrigation. For Wami farmers, this imposition of communal cultivation in the uplands and the abandonment of their fields in the lowlands was incomprehensible. It disrupted household food production and ran counter to who they were and what they practiced as "the people of the valley." Elders complained about numerous issues with the management of communal farms, including inadequate inputs, broken irrigation pumps, corruption, freeloading, and lack of clear guidance from district and village authorities on what people were supposed to do or how they were supposed to work together. It was not uncommon for people to withhold their labor in the communal farming scheme, elders said, and many felt discouraged as their hard work was not recognized. Throughout Operesheni Pwani, most farmers kept their options open for returning to their fields in the floodplains, and ultimately they did so when the campaign came to a close in 1976.[70] By refusing to work the way the state wanted them to and by reclaiming their lifeways, Wami farmers undermined the socialist experiment from the grassroots.

LARGE-SCALE CATTLE RANCHING

Looking at the spatial distribution of ujamaa villages registered in Bagamoyo District between 1973 and 1977, one notices a large gap in the coastal strip between Saadani, Matipwili, and Makurunge (figure 2.14). According to archival records, oral histories, and an interview with a longtime activist in Bagamoyo, this vast area of land—roughly thirty thousand hectares—was set aside as a gift from President Nyerere to the second president of Zanzibar, Aboud Jumbe, when the latter took office in 1972.[71] The first decade of the union between mainland Tanzania and Zanzibar had been marred by political unrest, with the first Zanzibari president, Abeid Karume, imposing numerous policies that challenged Nyerere's vision of national and African unity. Karume died unexpectedly in April 1972 in an assassination plotted by Abdulrahman Mohamed Babu, a Zanzibari pan-African nationalist and a major critic of the Karume regime.[72] Upon taking office, Jumbe implemented policies that aimed to relax tensions between the archipelago and the mainland, and during his tenure the Afro-Shirazi Party and the Tanganyika African National Union (TANU) merged to become the ruling party, Chama Cha Mapinduzi (CCM, the Party of the Revolution).

Elders in Wami recalled hearing rumors around the time of villagization that Nyerere was going to "gift" the land to Jumbe as a symbol of national unity and to recognize Zanzibar's commercial heyday in Bagamoyo in the late nineteenth century. According to one elder, the Zanzibari government originally set up a juvenile prison there but later decided to convert the land to a state cattle ranch:

> Way back then, it was like a prison. The [Zanzibari] government sent juvenile convicts [*vijana wahalifu*] here at first, but later turned it into a place for cattle keeping. The process of creating RAZABA started in 1976, but it was not until a few years later that it formally opened. I heard that Nyerere and Jumbe had some disagreements on how best to use the land. And then there were people in Winde [a historic coastal settlement] who wanted neither the prison nor the ranch.[73]

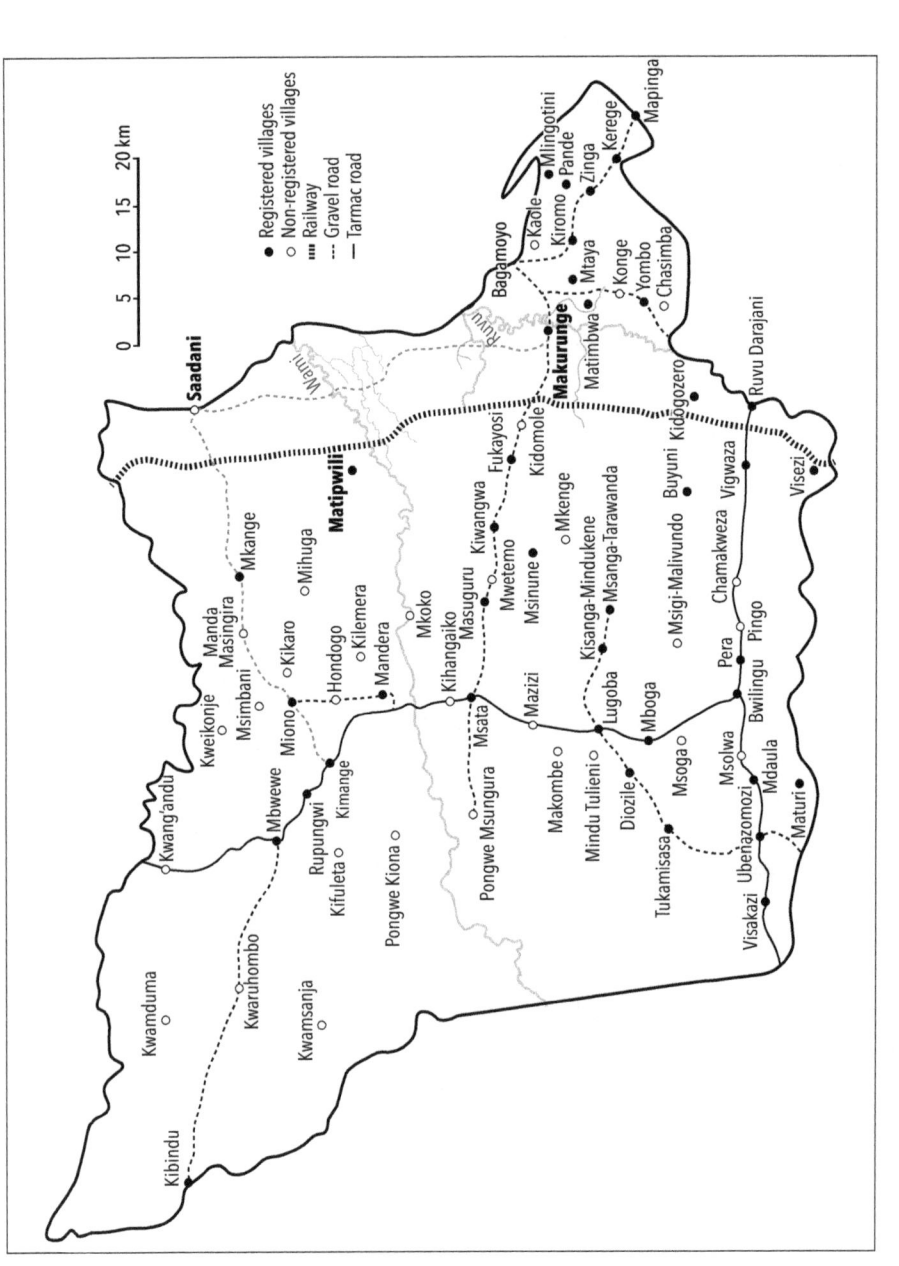

FIGURE 2.14. Distribution of *ujamaa* villages in Bagamoyo district as of 1978, with Saadani, Matipwili, and Makurunge in bold; no villages have been registered in the coastal strip between those three places. Map created by the author, adapted from Sitari, "Settlement Changes in the Bagamoyo District," figs. 3, 4, and 97.

The idea of establishing a cattle ranch in Bagamoyo coincided with the opening of the National Ranching Corporation in 1975. Along with the National Agriculture and Food Corporation established in 1969, these parastatal establishments embodied the second pillar of Nyerere's vision for rural development: large-scale plantations and ranches. On February 21, 1977, the Office of the Director of Land Development Services in Dar es Salaam (Viwanja–Dar es Salaam) sent a telegram to the Ministry of Agriculture in Zanzibar (Kilimo-Zanzibar) informing them of the transfer to the new cattle ranch of 77,663 acres (31,429 hectares) in Bagamoyo.[74] The telegram outlined the conditions of the transfer, including the need to demarcate clear boundaries and to ensure that the existing properties and the rights of people living inside the area were not to be disturbed in any way.[75] In February 1978, the Office of the Prime Minister and the Vice President sent a letter to the RAZABA manager confirming the boundaries of the ranch as follows:

South: Makurunge village
West: Tanga railway
North: Five Brake No. 10 up to the first junction where the river coming from the direction of the ocean bifurcates into two streams
East: Indian Ocean

Though local elders did not dispute the southern and western borders, they contested the northern and eastern boundaries. In an interview, the former ranch manager based in Bagamoyo, who assumed his position in 1985, admitted that he was unsure what "Five Brake No. 10" meant. Nevertheless, he was adamant that the Wami River was the ranch's northernmost boundary, though the ranch never reached that far in reality. Showing me two maps of the ranch, dated 1985 and 2008, he explained:

> We intended to use these areas [the floodplains on the south side of the river] for livestock keeping, as you can see from the maps. But we were using the land in pieces, I can say. We didn't utilize the land fully up to the river, but we didn't allow people there to stay either. Those areas were not part of any village. They were staying there illegally. The only people who were inside the ranch were those in Winde, and we compensated them when the ranch opened.[76]

The former RAZABA workers and Wami farmers I interviewed confirmed that, in practice, the ranch only reached up to the two livestock watering dams the ranch workers built, known as No. 4 and No. 5, about seven kilometers south of the Wami River. Elders in Wami dismissed the manager's claim, noting that neither the mainland nor Zanzibari governments consulted them about the land transfer and that it would have been absurd to keep animals in the floodplains.

Elders also disputed the manager's claim that the ranch compensated Winde settlers. Winde, like Bagamoyo and Saadani, was a key coastal entrepôt during the nineteenth century. A number of elders described how Wami farmers had maintained strong trade relationships with Winde, exchanging maize, rice, and vegetables for saltwater fish, copra, and palm leaves.[77] Between the late 1970s and early 1980s, with increased pressure and encroachment from RAZABA, several Winde settlers moved to Matipwili village, while others joined Makurunge or moved to Bagamoyo town. Still others remained on the land and even took RAZABA to court in 1989 after the aforementioned land rights activist in Bagamoyo introduced them to a lawyer who took on their case.[78] The High Court ruled in favor of Winde settlers and required the ranch to return the land to the people and pay them appropriate compensation. RAZABA has yet to comply with this decision. In 1994 Jakaya Kikwete, then the minister for water, energy, and mineral resources and MP of Bagamoyo, urged the government to take action on the Winde case. In his letter to the prime minister dated January 3, 1994, Kikwete argued that RAZABA's refusal to execute the court's decision was a "nuisance" to the citizens, and that it would be "harmful and scandalous" if they went back to the court to demand the judgment be enforced.[79] Recalling the advocacy efforts she was involved in with regard to the Winde case in the early 1990s, the activist described how politically sensitive the RAZABA issue was back then. It was on the eve of the nation's first multiparty elections, and tensions had been growing between the CCM and the new opposition party in the archipelago, the Civic United Front, founded in 1992: "You couldn't talk negatively about RAZABA, otherwise you faced political repercussions. The district commissioner, a CCM politician, told me then, 'If you touch RAZABA, the union government ends.' There was no room for discussion, even though there was no actual transfer of title between the mainland and Zanzibari governments for RAZABA. And sadly, Winde people never followed up with the case after that."[80]

The ranch operations officially began in 1979. Some of the ranch workers were recruited from local villages, but many others, especially livestock keepers, were brought in from central Tanzania, where animal husbandry is practiced widely. According to former ranch workers, RAZABA preferred to hire the Gogo agropastoralists from Dodoma with the assumption that "coastal people didn't know how to keep animals."[81] The number of workers ranged from three hundred to five hundred at a time, and the ranch housed as many as seven thousand cattle during its operative years.[82]

In the early 1980s, the workers asked the ranch manager for permission to set up their own farm plots within RAZABA. The manager agreed, and this led to the creation of a settlement in the southeastern part of the ranch, which became known as Bozi.[83] There, the ranch workers and their families could farm up to six acres, but they were restricted from planting trees to prevent future claims to landownership. According to former ranch workers, the wage they received was paltry, or, as they put it, "just enough money to buy salt."[84] As the men worked on the ranch, the responsibility for household food production and social reproduction fell directly on the women. The widows of former ranch workers described how they depended on their husbands to bring cow manure from the ranch to fertilize their crop fields, on which they grew maize, pigeon peas, vegetables, and rice, drawing water from seasonal ponds and streams during the long rains. Women also worked for independent income by selling food and home brews to the ranch workers. Soon, Bozi became a vibrant community. In 1989, a dispensary was established within RAZABA; it served not only the ranch workers and their families, but also the surrounding villages like Makurunge and Kidomole. During the same year, Bozi residents successfully lobbied the district government to open a primary school for their children within the ranch (figures 2.15 and 2.16).

By the early 1990s, however, the ranch was struggling to stay afloat. As the former ranch workers recounted, the last of the livestock were shipped off to Zanzibar in 1993, and all workers were dismissed by 1994; all other ranch assets were sold, and the dispensary, too, was closed by the end of 1995. Various theories exist about why the ranch closed, including insufficient veterinary services and a tsetse fly infestation. But according to the ranch manager, the fundamental

FIGURE 2.15. Razaba primary school, established in 1989. The school was forced to close in 2014 as many teachers left as a result of the uncertainties surrounding the EcoEnergy project. Photo by Zainab, December 2015.

FIGURE 2.16. Despite the Razaba primary school's closure, its grounds still serve as a meeting place for the villagers. Photo by Zainab, December 2015.

reason was the government's increasing indebtedness and the demands of structural adjustment. "A lot of state ranches and farms were closing at the time," he said. "The ranch faced economic difficulties, and the government didn't have enough money to run the ranch."[85] Unlike other state-owned farms at the time, RAZABA was never privatized. Seeing that the land was going to remain unused, the laid-off workers requested permission to remain there. The manager agreed, but under the condition that they continue cultivating only short-term crops and that they be ready to vacate the land when required. As he explained, "There was nothing written about the arrangement, just a verbal agreement. We told them to stay there as a means of earning their lives; they can cultivate annual crops like maize or something like that while they waited for other development to come. They knew they would be chased out someday."[86]

Yet, after the ranch remained inactive for five years, people began to make more long-term investments. They planted trees, such as cashews, mangos, papayas, and tamarinds, and built permanent houses. With no sanctions from the ranch manager, they felt justified in staying put and continuing on with their lives. The son of a former ranch worker who remained on the land after his father's death asked rhetorically, "Did RAZABA have the right to stop our lives just because it stopped working?"[87] The residents felt their presence further legitimated when the government officially registered Razaba as a subvillage of Makurunge in 1993 in advance of the general election in 1995. Since then, settlements within Razaba proliferated beyond Bozi, most of them concentrated around Wami's tributaries and the ranch's abandoned dams. Most of the farmers who migrated to the area after the ranch's closure obtained land through the customary method of clearing bush or claiming unoccupied area. The late 1990s also saw an increase in the number of pastoralists to the area, especially the Barabaig, many of whom were descendants of those who had been displaced from their homelands in north central Tanzania by a Canadian-sponsored wheat scheme.[88]

The closure of the ranch coincided not only with neoliberal economic and political reforms but also with the national land law reform. Widespread dissatisfaction among the rural population with villagization, together with growing donor pressures to promote a "free market" in land, spurred the process.[89] The reforms resulted in the adoption of the National Land Policy of 1995 and the enactment of the Land Acts of 1999: the Land Act (No. 4) and the Village Land Act (No. 5). While these new laws retained all land under state ownership, vested in the president, they nonetheless gave formal recognition to customary rights of occupancy.[90] Specifically, the Village Land Act provided for customary ownership not only to those who possessed land titles or certificates of customary rights of occupancy (CCROs) but also to those who could provide evidence of "peaceable, open, and uninterrupted occupation of village land under customary law for not less than twelve years."[91] Under this provision, the residents of Razaba who have been on the land for more than twelve years since the closure of the ranch would, in theory, be able to claim their customary land rights, although, as later chapters will show, the state would continue to treat them as squatters or intruders on government land.

WILDLIFE CONSERVATION

According to the Tanzania National Parks Authority (TANAPA), Saadani National Park covers eleven hundred square kilometers (110,000 hectares, or about 425 square miles) across three coastal districts, Bagamoyo, Handeni, and Pangani.[92] Prior to the national park's promulgation in 2005, a smaller portion of the park—approximately three hundred square kilometers (30,000 hectares), an area that would have overlapped with the LBS cotton plantation from the early 1900s—operated as Saadani Game Reserve.[93] Though the game reserve was officially declared in 1974, oral histories and secondary literature indicate that it had been operating since the late 1960s, when the amount of land under protection for conservation roughly doubled across Africa.[94]

Though some wildlife, such as the waterbuck, giraffe, lion, leopard, elephant, crocodile, and hippopotamus, are endemic to the Saadani ecosystem, other nonindigenous species like the oryx, ostrich, and zebra were imported to the game reserve between 1968 and 1974 from Arusha and Mbeya; all animals were kept in cages and fenced areas.[95] The game reserve's early visitors were foreign dignitaries who wished to "enjoy a break from Dar es Salaam and who came to hunt by special privilege."[96] Since independence, generating foreign currency through wildlife tourism has been a key motivating force behind Tanzania's expansion of protected areas.[97] Nyerere is quoted as saying around the time of independence, "I personally am not very interested in animals. . . . Nevertheless, I am entirely in favor of their survival. I believe that after diamonds and sisal, wild animals will provide Tanganyika with its greatest source of income. Thousands of Americans and Europeans have the strange urge to see these animals."[98]

Around 1968, the Wildlife Division, a state agency responsible for the management of game reserves, evicted a number of Tengwe-area settlers from their land without compensation.[99] Tengwe is situated about twenty kilometers north of the Wami River where the Link railway line intersects with the Mvave River, which flows eastward toward Saadani village before draining into the Indian Ocean. As significant as the Mvave was for Tengwe settlers as a water source, the Wildlife Division also coveted it as a potential dam site for the game reserve.[100] Once evicted, the displaced Tengwe residents resettled to Saadani and what later became Matipwili village. According to elders' recollections and the Wildlife Conservation (Game Reserves) Order, 1974 (Government Notice No. 265 and 275), the southern boundary of the game reserve was at a settlement called Maguko, approximately two kilometers north of Kisauke. As one elder explained, "Just north of Kisauke in Maguko is the boundary between Matipwili village and game reserve; there is a big hill there, so once you come down from that hill there is the border."[101] Throughout the 1970s, however, the game reserve began encroaching on village lands far south of Maguko. One elder recounted, "They did it little by little. First around 1968, then 1971, 1974, 1986, and 1999, until they took over Kisauke in 2003. We don't know the amount of land they took. It was all done by force."[102]

Throughout the late 1970s, it became clear that the game reserve was too small to house the number of animals kept in captivity. Financial constraints also made it difficult to ensure adequate access to food, water, and shelter for the animals. The zoo eventually closed in 1977; all herbivores were released, and carnivores were sold overseas.[103] A number of elders described how they began to experience greater crop damage beginning around this time, owing to wildlife intrusion. Female elders in particular complained how game rangers harassed them when they tried to access tree resources outside the reserve.

Efforts to revive the game reserve and involve local villages in conservation began in 1996, when the Ministry of Natural Resources and Tourism requested assistance from the German government. In 1998, Tanzania's Wildlife Division, in collaboration with GTZ, established the Saadani Conservation and Development Programme, and the following year a proposal was made to upgrade the game reserve to a national park. From the standpoint of the Tanzanian government, granting the area the highest level of legal protection was arguably necessary not simply for the sake of nature conservation but to generate much-needed foreign currency in the wake of structural adjustment and increasing debt dependency.[104] As a 2000 GTZ report stated, the national park had an "enormous potential to generate revenue," evinced by the increasing number of investors applying for licenses to develop tourism facilities in the area.[105]

The extent to which local communities were involved in the planning process for the national park is unclear, based on my review of over a dozen secondary studies the GTZ produced between the late 1990s and early 2000s. None of the Matipwili and Kitame villagers recalled consultations, though the rumor at the time was that the national park was going to be "owned by the Germans."[106] Archival records obtained from the Matipwili village office indicate that two former village leaders had in fact been involved in a reconnaissance survey that the district and game reserve officials conducted in 1999. The aim of the exercise was to recover the original boundaries of the game reserve vis-à-vis Matipwili village in order to propose new expanded boundaries for the national park.[107] The reconnaissance ended with a proposal that the southern boundary of the game reserve be extended to include Kisauke subvillage and other common lands.

According to former Kisauke residents, the district and village leaders did not consult them in the drafting of the proposal. The proposal itself was vague in its wording. Beyond incorporating Kisauke, it stated that the boundary expansion would start from the southern part of the reserve "from Mtoa Ngoma [sic] to Kaburi Wazi and Wami River confluence, leaving Matipwili village on the Western side."[108] Recalling the debates that took place during this time, one elder said regretfully, "It was clear TANAPA was tricking us with words. No one likes these things—having your land taken. But what could we do? Serikali ni serikali. We couldn't oppose the government."[109] Other elders echoed his sentiment even as they struggled to comprehend the rationale for the national park. As one female elder asked incisively, "What is the advantage of saving wildlife if doing so displaces people? Is protecting animals more important than protecting citizens?"[110]

Eventually, Kisauke residents acquiesced to the boundary expansion with the understanding that TANAPA would compensate them for their land loss and that they would maintain their right of access to common resources like Mto wa Ngoma. The male elder who had participated in the reconnaissance recalled that the original proposal was to include a portion of the stream so that the villagers could still benefit from it: "The boundary was supposed to be *near* Mto wa Ngoma, *not* the entire area."[111] However, just as the Wildlife Division had done in the past with the game reserve, TANAPA continued to encroach upon village lands beyond Mto wa Ngoma and Kisauke, to the south side of the river all the

FIGURE 2.17. A sign for Saadani National Park is affixed to a tree in Matipwili. Photo by Mwajuma, March 2016.

way to Gama. As Mwajuma, a female elder introduced earlier in the chapter, complained, "TANAPA has a habit of eating [*kula*] people's land and shifting their boundaries. They use different stones to mark their borders. Other times they just put up a sign on a tree [figure 2.17] and say, 'This is national park land,' when in fact we are the ones who have been using the land for generations!"[112]

In 2001, at a meeting held in Bagamoyo town, the proposal for establishing Saadani National Park was unanimously approved by TANAPA leaders and regional, district, and village-level officials, including the former RAZABA manager, the Matipwili village chairman, and the village executive officer.[113] Few affected villagers were aware of this event, nor did they hear any updates on the proposal until 2003 when they were forced to leave the land with forty-eight hours' notice. No public record of compensation payments exists at the village or district level. Many villagers believe that TANAPA gave the two former Kisauke subvillage chairmen a sum close to TZS 10 million—privately in the middle of the night—to coerce their fellow villagers to move (I return to these allegations in chapter 6).[114] In response to growing grievances among the displaced villagers, the district government made an effort in late 2004 to formally start the compensation valuation process in collaboration with Saadani authorities. Yet the process was halted even before it could begin "due to the complexity of the [the park's] southern boundary."[115] The district administration insisted that the boundary of the national park did not cross the Wami River, a position that the park leadership did not share. The former RAZABA manager also confirmed that the national park never crossed over to the south side of the Wami River.[116] While this debate ensued at the district level, the state formally declared Saadani National Park in 2005 through the Government Notice No. 281 under the National Parks Act (Cap. 228).[117] The notice indicates that the park's southern boundary extends to the south side of the Wami River. Various maps of the national park produced since 2003, including the tourists' maps found in the park offices, also depict the enclosure of a large portion of the north and south banks of the Wami River: the entire floodplain east of the railway line.

In 2011, to much local outrage, the national park established an office and a gate at the junction between the Makurunge–Gama road and the Wami River. Earlier that year, villagers from Gama, Kitame, and Matipwili had pooled resources to build a bridge that now lay beyond the gate. As an elder who was involved in the bridge construction recounted,

> It was the people of Matipwili, Gama, and Kitame who built that bridge. Before this bridge, we used to cross the river with a wooden pontoon boat which we pulled with a rope. One time the water level rose very high, and after all the mudflow, the banks slumped and created a steep slope. We needed a new way to cross the river. I was a member of the village council then. We had some budget in our village treasury, so we decided to use some of that money toward building the bridge. Most of us in Wami have family members on both sides of the river and in Bagamoyo, so it made sense to build a bridge. The bridge came first, before TANAPA built their office and gate there. So many people resisted. . . . And now TANAPA wants to take more of our land.[118]

The presence of the gate and the guards posted there meant that everyone who wished to cross the river, whether they were villagers or tourists, had to stop at the gate and pay the entrance fee to the park. The youth in Gama were the most incisive in their criticism of TANAPA. A young man in Gama whose extended family lived in Matipwili complained, "We don't like TANAPA. We don't want TANAPA. Why are they keeping us under their thumb? We can't even go through the gate without paying or bribing the guard. We once tried to put up a fight, and we were nearly beaten to death by the soldiers. They are not there to save the environment; they are there to eat [*kula*] the fees we pay." So deep were their distrust and contempt for TANAPA that villagers often described EcoEnergy in interviews as "TANAPA in disguise" or "just like TANAPA."[119] As one female farmer put it, "We are being squeezed on all sides by TANAPA and EcoEnergy. You know the saying, 'tembo wawili wapiganapo ziumiazo nyasi' [when two elephants jostle, it is the grass beneath that gets trampled]? We are the grass beneath."[120]

The villagers' confrontation with the national park continued to escalate over the years. In July 2015, when TANAPA sent a bulldozer to tear down trees in Matipwili to pave a new road for tourists, more than three hundred villagers joined in protest, with shovels and machetes in hand to block the move. In April 2016, village leaders wrote a letter to their MP, Ridhiwani Kikwete (the son of Jakaya Kikwete), to request that he voice their concerns to the Parliament; by the end of the year, they had yet to hear from him.

A Tingatinga illustration, a style of painting characterized by surreal and humorous caricatures and highly saturated colors, appears on the cover of Sekab BT's controversial ESIA, its environmental and social impact assessment, which the Tanzanian government approved in 2009. The artwork portrays the sugarcane estate, mill, briefcase-holding corporate/state elites, plantation and factory workers, wildlife, farmers, pastoralists, and fisherfolk all in seemingly harmonious coexistence. However, a closer look at the painting reveals several paradoxes. The (female) smallholder farmers shown, whom the sugar project presumably displaced, are only growing cash crops (pineapples) on marginal lands on the west side of the railway; they also face imminent conflicts with wildlife (a zebra is shown eating a pineapple plant) as well as with pastoralists and their livestock, whom the project also likely displaced. All freshwater sources are diverted to the sugarcane plantation and factory, and the wastewater flows directly into the Indian Ocean, where people are seen swimming and fishing. Meanwhile, the plantation also faces conflicts with the national park: a hippopotamus grazes on sugarcane as it inches toward the river for water, possibly polluted by nutrient runoff.

Clearly, EcoEnergy was not unaware of the contested land issues within its concession area. In fact, as chapter 1 discussed, these issues were flagged in the ESIA and were high on the list of debt financing conditionalities the donors imposed. According to Per Carstedt, when the company received an informal land offer from the government in 2007, it hired a lawyer to ensure that the promised land was in fact available and alienable for exclusive use. The boundary issue with the national park came up in the investigation, and the company expected the government to resolve the problem. Carstedt recounted in frustration,

> The lawyer looked into the matter and said there was ambiguity. Saadani National Park had expanded their borders in a very strange way. It was not approved by the minister of lands but the minister of natural resources. He recommended that we try to get more clarity before we make any more commitments. So we wrote a letter to the minister

of natural resources in 2008, and the minister's permanent secretary responded clearly, saying "No, no, we don't claim anything south of the river." We have that letter. That is the green light that we felt we needed to have before we started on our investments. . . . That was in 2008. Then, when we were talking with the AfDB, they had their own due diligence process, they had their legal experts. . . . They said, "When we are talking about investments of hundreds of millions of dollars, from a legal perspective, it is not enough just to have a letter from the ministry. It has to be more robust than that." Then they asked the government for the gazette [establishing the national park]. If you read the gazette, that area belongs to Saadani [National Park]. . . . Later, we received another letter from the permanent secretary in the Ministry of Lands clearly stating that they had agreed with the Ministry of Natural Resources to change the gazettement of that area [to make the land available for the sugar project]. This was in late 2013 [after EcoEnergy had already received the land title].[121]

His quote is telling and useful for summarizing the lessons of this chapter. Carstedt and other corporate actors attributed the project's impasse, among other things, to the complicated set of state actors and bureaucratic procedures involved in making land investible. This, of course, was not unimportant, but in abstracting land as property and a commodity to be exchanged, they neglected to understand the complexity and diversity of social lives lived on the land and the contested political-ecological history of the landscape. From the company's and donors' perspectives, as well as those of several government officials I interviewed, existing land disputes in Bagamoyo were largely contemporary and isolated issues that could be fixed through the application of techno-legal tools, such as rewriting government gazettes, issuing land titles, conducting surveys, and (re)drawing maps. These tools can conceal conflicts to some degree, but they fundamentally cannot erase the living practices, knowledges, and memories that have sustained rural social reproduction for generations, nor can they undo the long histories of struggle and suffering that have become deeply inscribed both in the landscape and in people's consciousness. Indeed, even with the possession of a land title, EcoEnergy could not transform the landscape into the harmonious, utopian scene the painting arguably represented. As partners in a joint venture, the company and the state still had to reckon with the hundreds of people who were living on and staking their claims to the land. As the next two chapters will show, controlling people would prove as important as controlling land.

3

ON BEING COUNTED
Gender, Property, and "the Family"

> **There are no boundaries for patriarchy.**
> Demere Kitunga and Marjorie Mbilinyi, "Rooting Transformative Feminist Struggles in Tanzania at Grassroots"

However tenuous and uncertain it grew over time, EcoEnergy's partnership with the Tanzanian government opened doors to new opportunities. Once the company aligned its business objectives with Tanzania's development priorities, it became eligible to apply for concessional loans and other financial resources from IFIs. This meant that the company could now borrow funds from organizations like the AfDB on more favorable terms than it could at commercial banks. But this unprecedented access came with a requirement to comply with international "best practices" or social safeguard standards on land acquisition and population displacement.[1] To benefit from financing support from the AfDB, the company, together with the Tanzanian state, had to demonstrate how they would avoid or properly manage forced displacement, a process of upheaval that donor agencies, for decades, have referred to with the vague euphemism "involuntary resettlement." IFIs define that term to mean situations in which "people are not in a position to refuse" or "do not have the right to refuse" land acquisitions that result in displacement.[2]

In the absence of in-house expertise, EcoEnergy hired an external development consultant to act as liaison with relevant government authorities to draft a Resettlement Action Plan in accordance with the IFI rules. As the custodian of all land in the nation, the Tanzanian government, via the Ministry of Lands, agreed to conduct a population census and property valuation, or what project planners referred to as the People and Property Count (PPC). The PPC, which began in late 2011, marked the first formal—and gendered—relationship the project forged with local residents, six years after its inception. Beyond simply offering a baseline for the Resettlement Action Plan, the PPC functioned as a de facto formalization process that rendered people and resources legible to the state. Studies of land formalization in Tanzania and elsewhere over the past two decades have drawn attention to the persistence of gender bias in the process, namely the privileging of husbands or male heads of households as the primary beneficiaries of land titles.[3] The PPC followed a similar pattern, but with a few exceptions, as I elaborate later in this chapter. Unlike other formalization programs, however, the PPC did not result in the distribution of land titles. Instead, the PPC resulted in other forms of documentation that nonetheless recognized certain individuals' (overwhelmingly men's) land rights, but with the direct intent of extinguishing those rights and paying compensation in return.

In this chapter, I consider the politics and limitations of the PPC, especially the gendered assumptions about "the family" and property that undergirded it. The story that emerges is a complex one. On the one hand, the PPC disadvantaged married women by subordinating their rights to land and compensation to the rights of their husbands. On the other hand, it recognized the rights of a minority of women, namely women "without men" on the margins of normalized conjugal and family relations. This latter group of women—comprising those who never married or who became single by virtue of divorce, separation, or the death of their spouses—were included in the PPC alongside and on equal footing to men. Though the documents they received were not land titles by any means, these women treated them as such and took pains to protect them. Yet they also realized the complex irony in all this: these documents were designed not to prevent but to effect legal dispossession.

Counting People and Property

The People and Property Count began with the short rains in mid-October 2011. Four teams led the exercise, with each team comprising a national land valuer, a district land surveyor, and two local field officers whom the foreign consultant recruited and trained. An early draft of the Resettlement Action Plan outlined various challenges that beset the PPC from the outset. Beyond general resource and logistical constraints, the draft report highlighted how the process was met with an "initial cold reception" as well as "saboteurs' activities from the impacted communities."[4] These references would later be omitted from the final version of the report posted on the AfDB and EcoEnergy websites. According to local residents, a certain group of male elders in Makaani had, indeed, incited people to refuse the PPC in defiance of the land deal; as I will return to in chapter 6, these elders later would go on to file a lawsuit against the government and EcoEnergy, resulting in a temporary injunction against the project. Given these complications, the scope of the PPC and the Resettlement Action Plan narrowed to a subset of households that would be impacted by the first phase of land clearance, mainly in the north. Though the PPC was partially resumed in March 2014, no addendum to the plan was ever published.

Most people did not outright reject the PPC as the male elders in Makaani did. But they recounted feeling extremely nervous and confused, because they did not fully understand what the PPC was or what it was for, let alone what the EcoEnergy project was about. Some people said they had heard about the EcoEnergy land deal from the radio and through word of mouth. Others had learned about it through a sign the district government had erected on the roadside on the route to Bagamoyo town. For many others, particularly women, elders, and those who lived in more remote areas with limited mobility, the PPC was their very first encounter with the project. Despite their apprehension, most people said they went along with the process anyway, because they felt they had no choice. As a sixty-one-year-old woman in Matipwili said, "I didn't know anything about the project until the government came [for the PPC]. They [the valuation and survey team] said our lives will not be affected by the project, because the investor will build us new houses. 'Why should I believe you?' I thought. But what could I do? The investor is working with the government, and small farmers like us can't fight the government."[5] Her neighbor, a man in his late seventies, was furious as he recalled the day of the PPC:

> They came just as I was leaving the house to go to the mosque. I got angry and chased them away because they had come without prior notice. Sometime later, they came back and held a public meeting with the villagers. They told us then that the government needed our land for investment. After that, I let them do their work because I felt like I had no choice. It was either I accept the project and receive compensation or refuse and be left with nothing. I was very troubled by this. I lived here for more than seventy years. The last thing they did before they left was to take my picture. I have not heard from them since.[6]

When I shared these reactions with a senior Ministry of Lands official, he was sympathetic at first but soon defended the government's position, which echoed the IFI definition of involuntary resettlement: "When the president has already decided that the project is for public purpose, people cannot refuse the project. It would be completely against the law. They just can't do that. There can be disputes about compensation, delayed or insufficient compensation, but people cannot dispute whether they want the project or not."[7]

As we shall see next, defining compensation, who deserves compensation, and how it is valued, was as murky as defining what counts as public purpose.

Defining Compensability

PAPs, or Project Affected Persons, was the generic term the project assigned to local residents. The PAPs were further categorized into those who were deemed "eligible" and those "noneligible" for compensation. There were differences, though, in the ways the Tanzanian government and the AfDB defined the parameters of compensability. While no separate policy on involuntary resettlement exists in Tanzania, the nation's constitution and various land laws and regulations stipulate that persons whose land is expropriated as a result of compulsory land acquisition must be paid full, fair, and prompt compensation within six months of land acquisition.[8] The Land Acquisition Act of 1967, enacted during the height of ujamaa, had initially limited compensation to the value of unexhausted improvements on the land, such as permanent dwellings, structures, and perennial crops like fruit trees. During the 1990s, the land law reforms and the enactment of the Land Act of 1999 in particular expanded the definition of compensation to include the commercial value of land, in addition to unexhausted improvements, as well as additional allowances for disturbance, transportation, accommodation, and loss of profits—all of which would be calculated and paid according to market rate.[9] While government regulations allow for in-kind compensation, cash payments have tended to be the norm.[10]

According to these laws, eligibility for compensation is determined by evidence of landownership. Few local residents possessed land titles at the time of the PPC, but as the previous chapter noted, under the Village Land Act of 1999, they could still claim customary rights by virtue of having cleared, occupied, and used the land consecutively and without interruption for twelve years or longer.[11] Despite this provision, data on land were not consistently collected during the PPC. The default response of government authorities I spoke with was that people in Bagamoyo did not deserve compensation, because they were illegally squatting on public land. As the chief valuer in the Ministry of Lands told me in an interview, "According to our laws, intruders are not supposed to be compensated. But because we had to follow international rules [with different eligibility standards], they got lucky."[12] The Bagamoyo district commissioner echoed a similar sentiment in an interview with the *Citizen*, saying that while the current occupants were not considered eligible for compensation in Tanzania, they would nonetheless be paid "compensation of compassion," in line with international human rights standards.[13]

In contrast to national regulations, the AfDB and other IFIs do not determine eligibility for compensation based on legal rights to land. That is, as long as people have been occupying and using the land prior to a project-specified cutoff date, which for the EcoEnergy project was the first day of the PPC, then they were considered eligible for compensation regardless of their status. Whereas the Tanzanian government calculates compensation at market value, international guidelines recommend that individuals be compensated for their losses at full replacement cost without depreciation, and that they be offered various in-kind resettlement assistance, examples of which I discuss in the next chapter. If the "spirit of compensation" in Tanzania, according to Wilbard Kombe, is to ensure that people "neither lose nor gain," the developmental mandate of IFIs requires that compensation restore and *improve* people's livelihoods and living conditions, before, during, and after displacement.[14] To bridge the gap between these two standards, EcoEnergy agreed to top up the difference between the government-calculated compensation amount and that calculated to meet the international standards.[15]

The PPC lasted for about a month, and individual households received at most two visits from the valuation and survey team. Although the event itself was short-lived, its effects were far-reaching. It generated a shift in the way people perceived themselves and those around them. If the project distinguished between eligible and noneligible PAPs, people began making similar distinctions among themselves between those who were "counted" and "not counted" by the PPC.[16] As a tool of governance and discursive practice, the PPC had produced that which it had named. It not only created subjects whose lives were now inextricably bound up with the project but also produced a new idiom of belonging based on difference and exclusion. As I elaborate in later chapters, people would draw on their status of being among the counted as a form of identity, as a weapon against the government's labeling of local residents as intruders, and as a way for them to distinguish themselves from landless migrants who moved to the area after the project's official cutoff date. For the purposes of this chapter, I focus my analysis on the role of gender, family, and marital status that decidedly introduced a wedge between the "counted" and the "uncounted."

The Natural Family

The government's draft valuation report comprised hundreds of pages printed in color, bound on the short side by a large black-plastic spine comb. I scanned through the pages of the report, albeit not in its entirety, in the presence of an anonymous source who granted me rare access to the document. The report contained information about the individuals who were registered during the PPC in 2011, including their names, location, an inventory of assets, and the total estimated value of compensation in monetary terms.[17] Each name in the report was accompanied by the individual's photograph, akin to a police mug shot, as the subjects held a small chalkboard sign in front of their chests with their unique reference numbers. As I turned page after page, it became apparent that records of women in the report were few and far between. Of the approximately five hundred names and photographs I counted by hand, more than 80 percent were male.

Few women I interviewed reported having been counted during the PPC independently of their husbands or male kin. Those who did were almost invariably widows, divorced, or separated women or unmarried single mothers, although a handful of married women happened to be counted because their husbands were absent at the time of the PPC. What distinguished these "women without men" from married women was that they possessed Land Form 69a signed in their own names. This was an official document people received during the PPC, establishing their rights as landowners and compensation claimants.[18] Another document people received was Valuation Form 1, a listing of all buildings, crops, and land that the valuers considered to be individually owned and deemed compensable to the extent that they had market value. Common property resources, which people depend on for food, fuel, fiber, medicine, building material, and cultural rituals like mkole (discussed in chapter 2) but which are not exchanged at the market, were excluded.[19] Although all compensation claimants should have received both forms during the PPC, the practice was inconsistent, especially with respect to the latter form. Some people said they never received it; some said government authorities promised to return with the form at a later date but never did; others noted that nothing was recorded in the form because they had no permanent crops at the time of the PPC and because they were told that the land belonged to the government. What remained consistent in people's testimonies was that whatever forms they had were predominantly in the hands—and names—of men.

The chief valuer I quoted earlier, who was also the only female government official I met during my research, explained the rationale behind this male bias: "Land Form 69 and the Val [sic] Form 1 are given to whoever owns the house. And you know in Tanzania the owner of the house and the head of the family is usually a male."[20] When I asked the official whether and how the government accounted for variations in the family form and changes to the family structure and composition over time (e.g., separation, divorce, death of a spouse), she demurred for a moment but responded by defending the public-private divide: "What happens in the family is not government's business.... Whatever the case, the assumption is that the man will manage the compensation in the best interest of his family."[21] Other officials I interviewed confirmed that in land administration and planning, the unwritten rule was to consider the nuclear family, presided over by a single male, as the basic unit of society. The idea of the patriarchal nuclear family, which both colonial and postcolonial states had naturalized to serve the needs of capital accumulation and nation building (see introduction), had become institutionalized into bureaucratic routines. As one senior Ministry of Lands official suggested, the primacy afforded to the male head of the household was both a result of bureaucratic inertia and the lack of clear procedures on how to account for difference:

> Honestly speaking, the issue of gender is still foreign in our government. For example, when I was involved in land titling ten years ago, I did not think about including a column for gender in the database— only names, parcels, and some additional attributes. No one asked me to do it, and it wasn't necessary because those who received titles were mostly men, unless a woman lived without a man. But I think there is an increasing need to do that nowadays, whether it is for land titling, property valuation, etc. How many men, how many women, how many men and women own land together.... Generally, though, I would say the practice has been inconsistent, and it's easier for people to follow what's already been done.[22]

From the married women's perspective, it was not all that unusual for the state to consider their husbands as the primary beneficiaries of the PPC. As they explained, husbands were expected to handle political and economic affairs, while wives were responsible for homemaking and care work. Even as they normalized these binary gender roles, just as the chief valuer had done, they expressed concerns about their ability to share in the compensation yet to come. Beyond the uncertainties surrounding the timing, amount, and type of compensation, they worried that their husbands, at some point, would misuse the payment. Several women reported that their husbands were using the anticipated cash compensation as bride-wealth credit to acquire new wives. Men were already misappropriating the *idea* of compensation, and this made their wives nervous about what might happen if and when compensation actually came.[23]

During the PPC, the foreign consultant recommended that all eligible people take in-kind compensation, such as land and housing, as opposed to cash.[24] Scholars of development-induced displacement have long argued that cash compensation can lead to greater impoverishment for individuals and families, with disproportionate effects on widows and older women.[25] Likewise in Tanzania, compensation for compulsory land acquisition has historically been paltry, especially in rural areas, and the state has consistently failed to generate sufficient funds to pay compensation.[26] As discussed in chapter 1, while the Land Act of 1999 was supposed to establish a land compensation fund, not a single government official I met could confirm its existence. Despite this track record and contrary to the consultant's advice, not all women I interviewed were opposed to the idea of cash compensation, and understandably. It was too early to decide, because they knew little to nothing about how much compensation they would receive, or what kind of alternative land they would be given in lieu of cash. Although they had their own views and preferences about the kind of compensation they desired, most married women concluded that the decision was not theirs to make. Consider the following excerpts from two interviews. The first was with a young woman in a monogamous marriage, and the second was with a middle-aged woman and first wife in a polygynous marriage.

> Woman 1: My husband will probably want to take cash compensation.... Money is like flowers; it does not last long. We don't even know how much we are going to get! I think land is much better than cash. If you don't have land, you don't have anything, although we still don't know kind of land we will be given. If we receive land that is bad, then it might be better to take cash.[27]

> Woman 2: Since men were the ones who were counted, it will be them who will collect the compensation. I have no clue how compensation will be paid, maybe through a bank account, cash handouts, ready-made houses, building materials, or land.... I trust that my husband will make a good decision, but sometimes you can't know for sure. I can't read his mind, what he is thinking or planning to do, which wives and children he will prioritize. If he takes cash, I will have to find a way to ensure that the money goes toward our children.[28]

In defending the naturalness of the nuclear family, government officials often invoked the neoclassical assumption of the household as a site of equity, sharing, and utility maximization. As the chief valuer stated, the husband was expected to act "in the best interest" of his wife and children in deciding how to dispose of the family income, including compensation for land loss. Yet this presumption of an altruistic household model flew in the face of how wives weighed their odds of benefiting from compensation against their perceived position in the family hierarchy vis-à-vis their husbands, and for some polygynous families, their co-wives. Conjugal and family relations thus were central to women's access to material resources. "The family," as married women lived it, was more a site of negotiation and bargaining than a site of equality and cooperation; or as Harriet Friedmann put it, the family is "a battleground over patriarchy where property is immediately at stake."[29]

Women without Men

Arguably, the People and Property Count reinforced patriarchal values both within the public and private spheres of life. It placed most married women at a disadvantage by making them legally, socially, and economically dependent on their husbands. On the flip side, however, the PPC offered an unprecedented opportunity for marginalized women "without men" to be recognized as independent landowners and compensation claimants. Such legal recognition was rare, and these women made sure to hold on to their land forms as a safety net against future dispossession and landlessness. In what follows, I present three cases of women "without men" of varying social positions, backgrounds, and spatial locations within the EcoEnergy concession and describe how they made sense of the land forms and their implications.

Amina

I first met Amina in August 2013. She was then a widow, fifty years old, who lived in Bozi with three young grandchildren. She was born and raised in Bagamoyo, as were her parents and grandparents. She belonged to the Nyamwezi ethnic group, whose traditional territories encompassed several regions in present-day west-central Tanzania. The Nyamwezi had first arrived in Bagamoyo during the late nineteenth century as caravan porters, forging long-term bonds with coastal peoples and contributing to the growth of Bagamoyo as a major terminus of the East African slave and ivory trade.[30]

Her father was one of the early workers at the RAZABA ranch, which opened in 1979, and subsequently her family became one of the first to settle in Bozi (see chapter 2). Amina remained in Bozi after marrying a fellow ranch worker, a Gogo agropastoralist from Dodoma, with whom she had six children. Until 2010, when her husband suddenly fell ill and died and when all her children had moved away, either after marriage or to find jobs in the city, she had been cultivating a six-acre plot, growing cereals, legumes, tubers, and other vegetables. Amina could only farm a tiny fraction of that now.

One December afternoon in 2015, I glimpsed Amina waving from a distance, her petite figure appearing and disappearing as I drove across the hills through the smoking dust of the dirt road. As I approached, I could see that she was not waving to greet me but gesturing for me to stop. When I pulled over and asked her what was wrong, she showed me pieces of a paper document she had carefully salvaged. It was Land Form 69a. Heavy downpours from the previous two days had caused her grass thatched roof to leak, flooding her entire house and damaging most of her possessions. She said she tried to save every piece of paper and dried them crisp under the sun in the hopes of putting them back together like pieces of a jigsaw puzzle.

The form was in no way a title deed, but she described it as such, referring to it as her *cheti* (certificate / land title). She asked me to help her put the pieces back together, and I did the best I could that evening with a pair of tweezers and wrapping tape. When I returned the form to her the next day, I asked why the paper meant so much to her. Her reply was unequivocal:

> Living alone makes you vulnerable, especially if you are a woman. Look at this roof. I can't even build myself a durable roof over my head. People take advantage of you or ignore you if you are a widow or if you are living without a *mzee* [an elder or respected person, in this case referring to a husband]. And living alone and waiting to be taken away from your land is worse than living with HIV/AIDS. But this form, it was given to me by the government, saying that *I* have the right to live here lawfully and that no one can evict me unless they compensate me. It proves that I have been counted and that I am the landowner.[31]

Rukia

I first met Rukia in August 2014. She was then a forty-one-year-old single mother living in Makaani with three children. She grew maize and cassava on a small plot of land but had recently started making charcoal for income, a gruesome, male-dominated job, which she described as "fever inducing." She belonged to the Ha ethnic group; she was born, raised, and spent most of her life in Kigoma in northwest Tanzania, near the Burundi border on the eastern shore of Lake Tanganyika. Most migrants I met within the EcoEnergy concession had come to Bagamoyo in search of a "coastal dream," to "find life" near the coast where there was perceived to be good land, water, and money. Rukia moved to the area in 2009 for similar but different reasons. She left Kigoma upon divorcing her husband; he had wanted to marry a second wife, and when she refused, he forced her out of their matrimonial home and prohibited her from using his clan land, which they had been cultivating together for over a decade. With no inheritance from her deceased parents and no natal home to return to, she decided to make the long journey across the country to Dar es Salaam with her children, in the hopes of finding work in the city as one of its many *mama ntilie* (female street-food vendors).

Life in Dar, however, was harsh. She could barely make enough money to pay rent, which she recalled as being TZS 10,000 (approximately USD 5) per month for one room. She and her children sometimes had to sleep on the streets. They eventually moved to Manzese, a slum on the outskirts of the city, where they squatted in an abandoned hut. One day she heard about a male elder in Makaani who was allegedly selling land to poor landless farmers (a story I return to in chapter 6). She paid the man TZS 5,000 and was granted permission to clear up to one hectare (about two and a half acres) in Makaani, though she and her eldest teenage daughter could only clear what she estimated to be a little more than an acre.

Rukia left a deep impression on me not only because of all the hardships she had endured in life, but also because of our unsettling first encounter. She had been sitting in a small open hut next to her house, feeding her children. When she saw me approaching, she got up suddenly and dashed into the house. She was in such a hurry that the *kanga* wrapped around her waist got caught on the door, which was made of uneven tree branches. She emerged shortly thereafter, holding a black plastic bag in her hand. She invited me inside the hut and offered me a plastic bucket to sit on. She untied the bag and handed me the piece of paper that she kept inside it. It was tightly folded in a tiny rectangle. I unfolded it carefully, straightening out the crease marks with my thumb while trying not to rip the fragile paper.

The document was Land Form 69a. She pointed to the line on the bottom of the form that read "Mmiliki wa Ardhi"—Landowner—under which she had printed and signed her name. She asked me to read the form out loud. Written in Swahili, the form stated that she had the right to claim compensation under national land laws, because her land and the improvements she has made on it were to be taken by the government for development purposes. When I finished reading the form and looked up, she asked me nervously: "So have you come to pay me compensation, or have you come to evict me?"[32]

Nuru

Nuru was a gregarious woman in her early sixties who lived in Matipwili with three unmarried adult children (all daughters) and one grandchild.[33] Her mother was a Kwere, born and raised in the Wami floodplains, and her father was a Bondei from Pangani, a district in Tanga region that abuts Bagamoyo District to the north. Her parents met in the 1930s when her father had migrated to Kisauke to work at the Wami Sisal Estate (see chapter 2). From there, they moved to Pangani, where her father was recruited to work at another colonial plantation. Nuru and her eight brothers were all born there. On the eve of Independence, her parents separated, and her mother returned to Matipwili with her youngest child and only daughter, Nuru, who was about seven years old at the time. Nuru remembered growing up in the floodplains, farming with her mother on one acre of land that belonged to her mother's maternal family. During ujamaa, when all villagers were forced to work in collective farms, they continued working the land for subsistence. Years later, even after her mother's passing, Nuru remained on the land. She never married but cohabited with a man with whom she shared the land and farmed for about ten years. After they separated, she lived with and bore children to two other men, who were now deceased and to whom she was also never married. When I asked her if she ever considered marriage, she responded with a story about her first partner, whom she referred to as her husband:

> One day he said he was going to help me collect fuelwood. I told him that was my job, but he insisted, so I let him do it. He was gone for a long time, and it was starting to get dark. I got worried, and so I took my machete and went into the forest to find him. When I got there, I found him with another woman. In shock, I yelled, "That man is my husband!" He got so mad and started beating me in front of the other woman. The other woman got scared and ran away. And after that I told him to get off my land. [*Pause*] Sometimes it's better to be without a man. If we had been married, it would have been difficult for us to get divorced, because I've seen husbands here bribe the village and ward officials and blame the wife.[34]

Unlike Amina and Rukia, Nuru had both Land Form 69a and Valuation Form 1. When I first met her in July 2014, she was one of the most outspoken critics of the EcoEnergy project. "Our land has been stolen. No amount of compensation would suffice," she had said then, showing me her forms.[35] When I visited her two years later, however, she had grown not only physically weaker but also more resigned. "The Nuru of those old days is gone," she said.

> I feel like an empty shell [*ganda tupu*]. I am feeling ill, and I can't keep up the fight. I can't farm as I used to. This life is hard. With whom and with what will I leave my children and grandchildren? Like today, I took the TZS 1,000 [less than 50 US cents] to the village to buy some paracetamol [acetaminophen]. It cost TZS 400, and then I bought some okra for TZS 300 and some eggplant for another TZS 300. By then I had no money left for sugar or salt. I told you before, I don't feel good about

being forced out of my own land as if we were wild pigs. But a stick from afar cannot kill a snake [fimbo ya mbali haiui nyoka].[36]

When I asked her what she meant by the proverb, Nuru motioned one of her daughters to bring the land forms from inside the house. Her forms were the same as the ones I had seen before, except they were now laminated, damage-proof. I couldn't help but think of Amina. She explained that she had been feeling intensifying pressure to give up her land, not only from EcoEnergy and the national park, but also from random land sharks from the city who targeted vulnerable villagers, including poor elderly women like her. Given these pressures, she wondered whether it would be better for her to concede the land to EcoEnergy than to risk being evicted with nothing. After all, no one else apart from EcoEnergy had promised compensation, and she had documentary evidence to prove her right to it. The land forms were the only tangible solution she felt she had to the threat of landlessness and impoverishment.

Amina, Rukia, and Nuru were accounted for in the PPC less because the government saw them as legitimate landowners and claimants than because, as officials openly put it, "there was no man of the house." They were single either by chance, such as through divorce or death of a spouse, or by choice, as in choosing not to be married. Whatever their circumstance, their stories dispel the assumed "naturalness" of the patriarchal nuclear family and offer a partial reminder of how diverse are the forms a family can take. If the PPC included these women "without men" by virtue of their deviation from the normative model of the male-headed nuclear family, it effectively excluded, and thus could render landless, married women in both monogamous and polygamous relationships.

In spite of the general apprehension people felt about the land deal, those "counted" in the PPC ultimately came to appreciate the fact that the state had legally recognized their land rights and rights to compensation for the seizure of their land. For those with fewer fallback options, like Amina, Rukia, and Nuru, the land forms offered a sense of security and autonomy. These women likened the documents to title deeds and insurance against unjust dispossession. They made every effort to preserve them in good condition, years after the completion of the PPC, even when it seemed doubtful whether the EcoEnergy project would actually happen. Amina sought help in reassembling her torn-apart form even before she could fix her leaky roof; Rukia kept the form tightly folded and hidden in a plastic bag lest it be stolen; and Nuru had her forms laminated to protect them from damage.

This particular finding raises some complex issues. Since neoliberal economic reforms, and especially after the publication of the Peruvian economist

Hernando de Soto's *Mystery of Capital* in 2000, many policy makers and donors have held the position that formalizing property rights is a solution to eradicating poverty and achieving economic development in rural Africa.[37] In Tanzania, since 2004, the World Bank and other donor agencies have poured in hundreds of millions of dollars toward land titling programs, with the aim of promoting tenure security and economic empowerment of rural people, especially women, and protecting them against unlawful land grabbing and ensuring transparency in large-scale land deals.[38]

A cursory reading of this chapter might lead one to conclude that including women in land formalization is both politically necessary and economically desirable. However, as I stressed from the outset, far from protecting people's land rights and their rights to remain on the land, the PPC and the documents that resulted from it were fundamentally tools to formalize dispossession. Having access to the land documents made Amina, Rukia, and Nuru feel safe to some degree, but it did not fundamentally change their marginal position in society as poor elderly widows, divorcées, and single mothers and grandmothers. They continued to feel "vulnerable," as if they were living in an "empty shell" and forced to engage in undesirable, "fever-inducing" work to make ends meet. Without addressing the larger socioeconomic system and political institutions that remain embedded within capitalist and patriarchal structures, land formalization can never be a panacea for gender inequality and land grabbing.

Furthermore, as the three women suggested in their narratives, the land forms were not valuable in and of themselves but only insofar as they could render possible the promised compensation. Yet so much was unknown about the details of compensation, much less the trajectory of the land deal. Compounding the uncertainty was the inability of the Tanzanian government and EcoEnergy to agree on who would pay for compensation and at what rate. On the one hand, the government pointed the finger at EcoEnergy for not offering to help make up for the shortfall in their compensation budget. As the chief valuer complained,

> EcoEnergy promised they would pay us an advance so that we could pay compensation. They said if we accept the advance, then our shares in the investment will be diluted later or something like that. That is what we had agreed on. For the first valuation, you know the PPC, we had some budget, but for the second, we had budget constraints. We had no money left to pay people compensation, so we were waiting to get funds from EcoEnergy. But the company has not done anything. Payment has been delayed, and it's demoralizing the people in Bagamoyo and also us in the government. I don't think the company was very serious about this issue of compensation and resettlement.[39]

On the other hand, from EcoEnergy's perspective, compensation was the responsibility of the Tanzanian state. As Per Carstedt reasoned,

> Actually, the government should deal with their own people. That's the right way. The same thing would happen in Sweden or Europe. If you build a railway or so, if you need people to be removed, then you would have to compensate those people so that they are at least as good or better off. Normally in Tanzania, people are worse off! The government thought following the international best practices for compensation was too costly. It set the bar too high for future projects. We tried to tell them, "You know, if you really want to develop and attract big institutions and serious business, then that is the minimum requirement."[40]

He acknowledged how frustrating the delayed project negotiations, and, by association, the delayed compensation, were becoming for many people. He even admitted that the PPC might have begun prematurely, although he concluded that there could have been no other option, given the pressure the company was facing from the AfDB at the time:

> If I were living there [in the project area], I would be frustrated, too, because I was told in 2011 that I would be compensated in 2012—well, because that's what the government had told us! Of course it creates all sorts of confusions when you send wrong signals. In hindsight, if we had known about all these issues with the land and the government, then we probably should have started the PPC later, when the time was right. But the banks were saying "this is social due diligence; you have to really start working on it [planning for involuntary resettlement]." So, it is a dilemma. I really, honestly, I don't see how we could have done things differently.[41]

In summary, though the subsequent steps did not occur, the PPC was an important mechanism for rendering technical the political and moral problems of displacement and compensation; as I return to in the conclusion of this book, the PPC and its uneven gender effects would have long-term ramifications in Bagamoyo. In the next chapter, I examine the consequent technologies of power the company and the state adopted to control both people and resources in the years following the PPC and amid growing uncertainty about the project's future. As I show, gender politics would remain central to governing this liminal land deal.

4

GOVERNING LIMINALITY
The Bio-necropolitics of Gender

> How is life weighted, disciplined into subjecthood, narrated into population, and fostered for living?
>
> Jasbir Puar, *Terrorist Assemblages*

"We see government cars passing, horns beeping, but we are told nothing," complained Mosi, a fifty-nine-year-old widow in Bozi. It was a dry afternoon in July 2014. We were sitting outside her house with a couple of her female friends who had gathered to weave mats from palm leaves. A government vehicle had just passed by, and the women followed the car with their eyes until it had gone past a hill. It had been nearly three years since the PPC, and anxiety ran high as people waited in vain for what was supposed to be imminent displacement.

"I feel bad waiting because I don't know what awaits me," said a younger woman sitting across from Mosi with a toddler on her lap.

"Maybe we'll all be dead by the time it comes," Mosi replied, and others chuckled. There was a grain of truth to every joke.[1]

Later that day when her friends had gone home, Mosi told me that she was less bothered by not knowing when the project might begin than not knowing what exactly the project was and who was responsible for it: "What is this project? Who is this project? They keep changing time after time. We don't know who's who, and they keep changing their words, too. First, there was Sekab and then EcoEnergy. Then there was a white woman/mother [*mama mzungu*], who gave us the trainings. And now we have these paramilitary forces [*mgambo*] who are beating people, stopping us from farming, taking our lives hostage. It's very confusing."[2]

Although negotiations remained stalled in Dar es Salaam, the project lived on in Bagamoyo in unexpected and discontinuous ways. As Mosi highlighted, the project unfolded not through the direct engagement of the principal actors behind the land deal, but through two distinct third-party entities. First, there was the foreign development consultant ("mama mzungu") whom EcoEnergy had hired to manage the PPC and the resettlement planning process, introduced in the previous chapter. Second, there were the civilian paramilitary forces (the "mgambo"), whom EcoEnergy contracted from the district government to secure the project site boundaries. The corporate executives remained at a distance in Dar es Salaam and managed the project through these intermediary actors as well as two Tanzanian employees based in Bagamoyo town.

In this chapter, I explore the role of the consultant and the paramilitary, their techniques of governance, and their gendered effects. Drawing on Michel Foucault and Achille Mbembe, I contrast the biopolitical governmentality of the former, which aimed to improve the life chances of to-be-displaced populations through alternative livelihood skills-training programs, with the necropolitical governmentality of the latter, which served to deny the possibility of agrarian life through land use restrictions, surveillance, threats, and physical violence. These distinct interventions stemmed, on the one hand, from EcoEnergy's need to adhere to international safeguards on involuntary resettlement and, on the other hand, from its need to assert its authority and property rights or the rights to exclude others from land access.

While seemingly contradictory and conflictual at first glance, both mechanisms of power were ultimately similar in their assumptions and effects. At the broadest level, they both presumed an ideological representation of the local people as an object of control or an "other"—lacking in capacities to choose or make decisions in ways that the project deemed desirable. At the same time, dominant gender ideologies that naturalized certain activities as feminine and masculine, or as socially acceptable ways of being and doing women and men, guided their everyday operations. Questions such as who could take part in a tailoring training, who was culpable for cutting down trees, and who should remain on the land amid increasing violence all hinged on the definition and negotiation of gender. And the answers to these questions, as I show, would have direct impacts on intrahousehold power relations and broader local struggles for survival and social reproduction.

Gendered Biopolitics of the "Early Measures"

Soon after completing the first round of the PPC in 2011, the foreign development consultant, whom I will call Anna, began implementing a pre-resettlement assistance program, which became known as the "Early Measures." The program involved providing a range of livelihood skills training to the PAPs, the "Project Affected Persons," so they could "seamlessly begin to support their families upon resettlement and work towards a more productive future."[3] The types of training offered included basic literacy and numeracy, catering, baking, tailoring, hairdressing, driving, welding, carpentry, information technology, mobile phone repair, motorbike repair, commercial horticulture, aquaculture, poultry keeping, and entrepreneurship. The project's Resettlement Action Plan described these training courses as being critical for "engendering self-reliance" and inducing "mind-set changes" among the local people about the possibility and desirability of life without land and land-based livelihoods.[4] From Anna's standpoint, introducing the Early Measures was an imperative, not an option, if the project wished to benefit from development financing. The AfDB, for instance, requires that involuntary resettlement be "conceived and executed as part of a *development program*," so that the "standards of living, income earning capacity, and production levels (of displaced and to-be-displaced populations) are improved."[5] Intentionally designed to "make live" and "improve life," the Early Measures was quintessentially a biopolitical program.[6] Yet despite its incitement to life, it was fundamentally an instrument for facilitating or lessening the blow of displacement and dispossession yet to come.

The Early Measures ran for approximately twenty-nine months between 2012 and 2015. During this period, EcoEnergy disbursed about USD 29,500 toward the program, a figure that paled in comparison to the total estimated cost of the investment project, valued at half a billion dollars.[7] With the limited budget she had, Anna contracted community-based organizations and technical colleges in Bagamoyo to provide the offered training. Some modules, such as basic literacy, numeracy, computer skills, and entrepreneurship, were offered "in-house" by young Tanzanian college graduates Anna hired as interns. Once training courses were completed, Anna offered microfinance loans to select participants—those who had attended all courses, passed final tests, received good reports from instructors, and completed a business plan—so they could "put their skills into practice."[8] The average loan and interest rate was TZS 200,000 (USD 100) and 5 percent respectively, and the interest payments collected were expected to subsidize future loans to other trainees.[9]

Anna was a white middle-aged British woman with an impressive portfolio of professional experiences in Africa spanning over two decades. Project documents referred to her as a "resettlement expert."[10] She and the chief valuer who appeared in the previous chapter were the only two women vis-à-vis over forty men in leadership positions I interviewed at EcoEnergy and across different levels of the Tanzanian government. Of all the state and corporate actors I interviewed, Anna was the only one who raised concerns about gender regarding the EcoEnergy project. Acknowledging that "development projects can sometimes bring more harm than good," she stressed that a careful participatory process and an assessment of the needs and aspirations of the PAPs underpinned the design of the Early Measures. For instance, she highlighted that training in areas such as hairdressing, catering, and tailoring were incorporated into the program specifically to increase the participation of women. Not only did these skills constitute work that was seen as culturally appropriate for women (as opposed to more "masculine" professions like carpentry and masonry), but also the female PAPs had expressed interest in them. Anna also underscored that unlike the male EcoEnergy executives who approached the PAPs largely from a business perspective, she intended to build lasting relationships with them. She said that during her earliest meetings with the PAPs she introduced herself as a "mother" to drive home the point that she was coming from a place of care, not profit. Though the program was financed by EcoEnergy, she highlighted that each training course was delivered "independent of the project going forward or not," so that the PAPs could still benefit from the skills they acquired.[11] While there was little doubt of her good intentions, her frequent use of the acronym "PAPs" in her narrative and also in project documents had the inadvertent effect of rendering the local people apolitical categories or powerless victims who needed help from outside "experts."

Among those who participated in the training courses, perceptions and assessments varied. In the early days of the program, several individuals expressed appreciation for the learning opportunity. One male participant in a carpentry course said, "I feel like I am no longer blind, as if I discovered a key to a locked door. I am learning something new."[12] Another young man who participated in a driving course said, "I don't know what will happen with the project, but I feel like I am building some hope with this drivers' education. It could be a starting point for something."[13] A middle-aged woman who participated in an entrepreneurship training course shared similar views: "I am very grateful. After the training, I received a loan to start a clothing business. I went to town and bought some *kanga* and *vitenge* [wax print fabrics], and I sold them in Gama and Kitame. I bought twenty-six pieces, and now I have six left. I received a loan of TZS 175,000 to continue the business."[14]

Others, however, expressed ambivalence and cynicism toward the Early Measures. Another young male carpentry participant remained skeptical: "There are two groups around here. One group thinks positively of the trainings and the loans. But there are others who don't feel good about them because they are connected to the EcoEnergy project. Some don't feel good because they didn't get a chance to participate in the trainings like some elders and women, and others just think the whole thing is some sort of propaganda by the investor."[15] One male participant in the entrepreneurship training said he was worried about the implications of taking the loan and the possibility of defaulting on it:

> I considered opening a business to sell rice but rejected the loan because the terms were not fair. I was offered TZS 270,000 and was told that the interest would be TZS 13,500. I was expected to repay the loan in two weeks' time, which was not enough time, especially for someone like me who has never run a business before. And it wasn't clear to me whether I would lose my compensation if I failed to pay back the loan.[16]

Few people I interviewed remembered Anna by her name. As Mosi described earlier, many people knew Anna as *yule mama mzungu*—that white woman/mother. Her gender and race—and the structural power and privilege she thereby wielded—served as intersecting markers of difference and a frame with which people made sense of her role and their relationship to her. Perhaps influenced by the way Anna introduced herself as a mother, people also described their relationship to her, and by extension to EcoEnergy, using the parent-child metaphor. Consider the comment by a woman who participated in a catering course: "We are treated like children; we are just expected to obey and accept whatever we are given. Training is not what we want. We want land. Land isn't the same thing as training. Say the project area later becomes a good place for doing business. But what if I lose my land and have to move somewhere far away where there isn't a market? How could I possibly use my catering skills then?"[17] An elderly man, who did not participate in any of the offered training, echoed her sentiment: "You know, when you want children to go and do something for you, you have to say to them, 'Go, I'll give you candy as a reward.'"[18] In his view, EcoEnergy, via Anna and her staff, was offering people sweets, in the form of skills training, to get them to acquiesce to the project and concede their land.

As one of the trainees suggested earlier, achieving inclusive participation in the Early Measures was a challenge, and Anna and her team recognized it as such. Many elders said they either heard about the training offerings too late, or they did not feel welcome to participate because of their age, or they failed to participate owing to poor health and limited mobility. Another barrier was the minimum education requirement for certain courses. For instance, a divorced woman who participated in an English literacy program noted that the training was available only for those who had finished Standard 7 (seventh grade); the teacher had to send many people back home, especially women who had not gone to school.[19] In addition to these obstacles, married women and unmarried daughters needed to seek approval from their husbands or fathers if they wished to enroll in any of the training courses. This was not only because of cultural norms, but also because most courses took place outside their homes and villages. The trainees were expected to reside in Bagamoyo town for the entire duration of their course (anywhere from two weeks to three months), with some occasional trips to Dar es Salaam and to other regions, depending on the kinds of training they were involved in.

For many husbands, the possibility of their wives leaving and withdrawing their labor from the farm and the home, albeit temporarily, posed not only a challenge to the daily reproductive needs of the family, but also a threat to their conjugal authority, namely their perceived right to claim control over their wives' labor power. A thirty-two-year-old man justified denying his wife's request to participate in tailoring training while noting he himself had enrolled in an English literacy program: "What men decide is the family decision. Women are needed in the home to take care of the family. She needs to cook, wash dishes, clean the house, fetch water, cut fuelwood, take care of children, and help the man with farmwork."[20] While women "without men" were not subject to this kind of male authority, those who had children or grandchildren nonetheless had to also negotiate the needs of care work if they wished to participate. In this way, the Early Measures program was deeply entwined with the domestic politics of social reproduction; women's participation in the training courses triggered intrahousehold conflicts, tensions, and negotiations over who was going to perform the unpaid care work, the essential labor of maintaining life on a daily basis. And at the crux of these intimate negotiations was the definition of gender and gender roles. This dynamic is best illustrated with the following vignette of a married couple, Zainab and Yusuf. Though their experiences are situated in their particular conjugal context, the themes that emerge from their testimonies resonate with patterns observed in interviews with other families.

Zainab and Yusuf

The first time I interviewed Zainab at her home in Bozi in August 2013, she seemed timid and guarded. I presumed this was because Yusuf, her husband, sat with us the entire time, answering my questions on her behalf. During this first interview, it was he who told me about the couple's "shared" experience of the Early Measures. He did not hold a favorable view of the program:

> I was trained to be a driver, and she, a tailor. We were told that we would be given two months' worth of training, but in fact each of our trainings was less than a month long. We didn't learn much. The trainings were not given to help us citizens out of poverty, but to show that EcoEnergy was doing something nice for the local people. Even those people who

are professionally trained for a year still struggle to make it, so you can imagine—what can we farmers do with less than a month of training? After training was over, EcoEnergy called the district commissioner to come and hand us our certificates [of completion]. There were photographers. They were just putting on a show.[21]

On my return trip to Bagamoyo in July 2014, I met Zainab again—not at her home but at her parents' home in a neighboring hamlet just outside the southern border of the EcoEnergy project site. She and her husband had gotten divorced two months earlier, and she was forced to return to her natal home. "EcoEnergy affected my marriage," she started by saying.[22] I asked her if she felt comfortable sharing. She continued:

> You know, my husband didn't want me to take part in the tailoring training. I told him maybe if I could learn how to sew and make clothes, I could bring more income to our family. It was then when he suggested the idea of marrying another woman. He asked, "Who will clean the house and feed the children when you are gone?" I said maybe we can arrange someone, be it our relative, someone from outside, or a friend, to do that work for me just for a few weeks. And that's what we had agreed on.
>
> But when I came back after the training, he said he still wanted to marry another girl. That girl was not older than sixteen. He said he had already gone to visit her parents to ask her hand in marriage. He said, "We need another woman in the house if you are going to go out and earn money making clothes for other people when the project starts." It pained me. I cried so much out of pain and anger. I told him, "We have two children together. What would they think?" We fought and fought until the day I could not endure anymore and asked him for a divorce. I told him that we should divide everything equally, especially the things we owned together as a married couple. But he said that will not be possible because everything we owned had been valued by the government, and only his name was registered during the valuation. He said I was not entitled to compensation, whether in terms of farmland, housing, or money. He said the compensation would be for our children. I told him I will not agree to that. [*Long pause*] I don't know what to do. According to our Muslim faith, we have ninety days after the divorce to make it final. If you decide to return to your husband within the ninety days, then the divorce can be annulled. I will have to decide what to do by the end of next month.[23]

When I saw Zainab again a year later, in September 2015, she had returned to her husband. She invited me over one Tuesday morning when Yusuf was out at the weekly market. She said she decided to get back together with him to provide stability for their children and to reclaim her stake in the future compensation payment. Her husband, in the meantime, had married the young girl and set up a second home in the village. Her plan, she said, was to acquiesce to her husband's desires for the time being and then find a way to take her share of compensation in the future and leave him for good. How exactly that would happen, she had yet to figure out. Quietly enduring an unhappy marriage and the unequal power relations that undergirded it was a prosaic but nonetheless hopeful way for Zainab to regain her personal dignity and (the promise of) economic autonomy. I observed a similar pattern among other women who were dissatisfied in their marriages but could not ask for a divorce from their husbands for fear of losing compensation, regardless of the uncertainties surrounding it.[24]

Whether Yusuf had been having an affair before EcoEnergy arrived is unclear and less important here than the fact that Zainab's participation in the tailoring training—and the possibility of her entry into the labor market—ignited their conjugal conflict and frictions over the demands of household social reproduction. The idea of Zainab not being able to fulfill what he assumed to be traditional feminine duties in the home encouraged Yusuf to seek another woman and her labor power to ensure his (free) access to food provision, child care, cleanliness in the home, and sexual pleasure. As a result, he was able to assert and maintain his dominant status as the "man of the house." The Early Measures, which intended to produce self-governing subjects, ultimately reproduced gendered subjects.

From Consent to Coercion

Other changes were perceptible in the landscape when I returned to Bagamoyo in September 2015. The most visible were the notice boards Anna's team had set up to share news about the Early Measures, which had now fallen into a state of neglect. The glass panels on some had been shattered; announcements pinned to the board had fallen off or become tattered; and photographs had faded and become discolored from extended exposure to sunlight. "This is what yule mama left behind," said Halima, a female farmer in Matipwili, describing a photograph she took of a notice board still standing next to her house in front of a wild fig tree (figure 4.1). "I don't remember the last time anyone came to update it. It hasn't been maintained for a long time."[25]

As examined in chapter 1, the financial viability of the EcoEnergy project came increasingly under threat throughout 2015. According to Anna, EcoEnergy

FIGURE 4.1. A neglected notice board displaying information about the Early Measures. Photo by Halima, July 2016.

unilaterally terminated her contract around mid-2015, which resulted in an immediate and indefinite suspension of the Early Measures. She said that EcoEnergy had failed to pay her invoices for nearly six months before the termination of the contract, during which time she had to meet all program expenses out of her pocket, including her employees' wages.[26] Once Anna and her team withdrew from all ongoing activities, the only remaining actors that represented EcoEnergy on the ground were the civilian paramilitary forces, the Jeshi la Mgambo (People's militia), commonly referred to as mgambo. Their presence was not new; EcoEnergy had deployed them in 2014 when the company temporarily began a series of early works, namely the installation of cement posts around the project site boundaries to close off roads and entryways and to prevent unauthorized movements of people and goods (see introduction).

The history of the almost exclusively male mgambo merits special attention. TANU, the principal political party behind the nation's struggle for independence, instituted the mgambo in 1965 as a volunteer force controlled exclusively by the party. The idea was to end the army's monopoly over military training of civilians and thus ensure that "the party was on an equal security footing with the state."[27] Together with the TANU Youth League (TYL), this newly formed reserve army, primarily comprising politicized young men, was expected to defend the nation against internal and external enemies of the time.[28] At the height of ujamaa, Julius Nyerere placed these young men of the TYL and mgambo on a pedestal as vanguards of the socialist revolution and protectors of national security, while upholding women as custodians of the home and guardians of family well-being.[29] Throughout the 1970s, the mgambo assumed a range of functions, including enforcing the implementation of ujamaa villagization in rural areas and patrolling urban spaces together with the police and recruits of the National Service (Jeshi la Kujenga Taifa, literally Nation-Building Army).[30] In 1975, the People's Militia Act conferred on the members of the mgambo powers of arrest and search equal to those of a police officer.[31] By the late 1980s, the mgambo had become known not for their nationalistic origins, but for their rampant abuse of power and extralegal coercion. As Issa Shivji writes, ordinary citizens were subject to "gruesome beating, torture, and death" by the young, militarized men who were "heavily armed, trigger-happy and reckless with their arms."[32] Together, the mgambo, the military, the police, and other law enforcement institutions created in the wake of decolonization came to signify, enact, and maintain what feminist political theorists and philosophers have called the "masculinism of the state": the ways in which socially constructed ideas of masculinity, especially the logic of masculinist protection (of population, property, and territory), shape and legitimate the state monopoly of force.[33] As I demonstrate below, this gendered, arbitrary exercise of state power is usefully elaborated with Achille Mbembe's notion of necropolitics, which he described as the intentional infliction of violence against certain civilian populations, such that it results in the constant "subjugation of life to the power of death."[34]

Gendered Necropolitics of Paramilitary Violence

At any given time, approximately fifteen to twenty members of the mgambo were deployed at various security checkpoints across the project site (figure 4.2). Their presence fluctuated, as the turnover rate was high, and each member would be rotated to a different post every now and then. Though their numbers were not many, their presence itself, including their appearance in military uniform, possession of weapons, and thuggish behavior, was enough to instill a sense of fear and trepidation among the local people.

One afternoon in December 2015, I had a rare opportunity to interview one of the new mgambo recruits at a checkpoint near the southern boundary of the project site. He was a young man in his early twenties. He had been deployed a month previously from a village in another part of the district, approximately seventy kilometers away from the project site. When I asked him how he got the job, he replied in a tone that bore a sense of shame: "I needed the money."[35] He said he phoned his friends and connections in the district for months until he landed this job. It was an admission that reflected the growing crisis of social

FIGURE 4.2. *Kituo cha ukaguzi* (inspection checkpoint) near the northwestern border of the project site. Photo by the author, June 2016.

reproduction and masculine identity for poor young men, a pattern observed not only in Bagamoyo but elsewhere in Tanzania and across urban and rural Africa.[36] Individuals like him who volunteer to join the mgambo today are still trained by the military and armed with weapons, but the principal motivations behind their enrollment have arguably less to do with their allegiance to the ruling party, its political ideology, or patriotic nationalism than their need for money and daily survival. When I asked the new recruit how he would describe the mgambo's relationship with EcoEnergy, he stated that he had not been aware of the company or the land deal until he arrived on-site. Nevertheless, he sounded confident in describing the duties that were assigned to him: "Our job is to protect the project area. The people here are intruders. I am told to make sure that they are not cutting trees for fuelwood and charcoal. When we find them doing these things, we turn them in to the police, so they can be prosecuted."[37]

When I asked the same question to the two Tanzanian male EcoEnergy staff based in Bagamoyo, Saidi and Ali, they, too, stressed that the primary role of the mgambo was to protect the company's private property. During our first meeting in March 2016, for instance, Saidi, who introduced himself as EcoEnergy's communications officer, brought me a copy of EcoEnergy's land title. To explain why the mgambo deployment was necessary, he flipped through the document until he got to the page where the terms and conditions of the land lease were outlined. He silently traced his finger across the page; when he got to the clause that he was looking for, he emphatically tapped on the table and said: "Okay, read this. The title says that we have to 'take all measures necessary to protect the soil, prevent soil erosion, occurring thereon or outside the boundaries of the Land. . . .'"[38] He accentuated the first six words and then trailed off, perhaps because the clause had been written in specific reference to EcoEnergy's agricultural activities, which had yet to begin.

One might argue that the deployment of the mgambo benefited EcoEnergy, as it allowed the company to strike three birds with one stone. First, the company could save costs by replacing the relatively expensive foreign consultant (Anna) with the mgambo, most of whom were unemployed poor young men from the district who were willing to work for low wages. Second, particularly as the company's relationship with the national state was weakening on the cusp of a regime change, it could take advantage of critical state powers and functions that were devolved to regional and local political authorities, including the maintenance of law and order through the use of armed forces.[39] And finally, by allowing the mgambo to occupy the concession area, the company could forcefully demonstrate its exclusive right to property vis-à-vis the local people, to whom entitlements to land emerged not necessarily from formal titles, but from occupation and use.

When Anna was the primary governing agent of EcoEnergy, her team did not police people's everyday behaviors and labor practices. If they saw people planting tree seedlings or building new structures, for instance, they simply reminded them that they would not be compensated for those additional items beyond what had already been recorded during the PPC. After Anna's departure, however, people began reporting increased surveillance, harassment, and violence by the mgambo. In addition to enforcing existing restrictions on planting perennial crops and building permanent housing, the mgambo sanctioned and randomly punished people for everyday dwelling activities, such as planting annual crops, growing vegetables, repairing homes, and gathering fuelwood. The mgambo's training had convinced them that their role was not to educate people on what they were or were not allowed to do on the land. Rather, they were in the business of coercing people to behave differently—of coercing submission by drawing on the masculinist logic of protection based on the mobilization of fear.[40] As the young mgambo recruit quoted above stated plainly, their primary role was to protect and patrol the property.

Threats, intimidation, and petty corruption were common repertoires of mgambo domination. Sometimes they would hunt wild rabbits and birds in the forest and force the local people to buy the game meat; if people resisted or pleaded that they did not have the money, they faced verbal abuse and threats of violence and incarceration. Other times, the mgambo would ask people to pay bribes as high as TZS 50,000 (USD 25) if they wanted to carry on with their activities. This was not an option for the vast majority of people, for whom it would take weeks or even months to raise that kind of money.

Women and men were equally subject to the destruction and theft of property by the mgambo, but when it came to physical assault, men were the primary targets, with a few exceptions. The victims of assault recounted traumatic experiences of being slapped, kicked, and beaten by the mgambo to a point where some experienced broken limbs. Some were forced to get down on all fours and do hundreds of push-ups. Others were ordered to crouch in a low squat position and do hundreds of bunny-hops until they could not feel their legs anymore. If they resisted or failed to do so, they received further corporal punishment or risked having their properties destroyed or looted at another time. All in all, people described the disparate but similarly coercive mgambo activities as *mateso* (suffering/torture). The words *vurugu* (confusion/disturbance) and *fujo* (chaos/mess), too, became common parlance.[41] In less than a year after the sudden discontinuation of the Early Measures, the landscape had turned into a militarized, masculinized, and necropolitical space, where the relations between life and death became viscerally felt.

One time, the mgambo allegedly dragged four people out of their homes at two in the morning, accusing them of making charcoal illegally on EcoEnergy's private property. Meena, a fifty-year-old woman in the settlement of No. 5, who was one among them, recalled the night:

> There were four of us. We were asked to line up. We saw that there were four buckets of water. They poured cold water over our heads. I was only tortured with water, but the men were beaten until blood came out of their bodies. The mgambo told them to do push-ups. They commanded the men to go down and up as they smacked them with a baton. Before I was dragged out of my house, I asked them, "Why should you come here at night? What have I done? Are we being evicted?" They told me not to ask questions but to just come out. I resisted, so they kicked and slapped me.[42]

The violence Meena experienced was exceptional. As noted earlier, while women experienced verbal harassment and intimation by the mgambo, physical violence against them was rare. Several women attributed this to their status as *watu wa nyumbani* (housewives, or literally "people of the home"). They speculated that the mgambo likely assumed that household decision making, including decisions over how to dispose of family land, was the prerogative of fathers or husbands, although the reality was not always that clear-cut. Presumptions of criminality were more attached to activities that were perceived as "men's work," such as land clearance and charcoal production.

Another reason for the disproportionate harm against men was that many women and children had left the area since violence escalated. Many feared that the land was becoming "no place for a woman," and the left-behind husbands and fathers suggested that they would invite their families back only if the land became habitable again.[43] The exodus of women and children was especially evident in the settlements in the northwestern part of the project site, where the largest number of migrant populations resided and where the mgambo maintained the heaviest presence. In a photo-narrative, a young male farmer describes one particularly intense confrontation the residents of this area had with the mgambo and the district police commissioner in December 2015 (figure 4.3): "The mgambo have been intimidating the villagers. They have taken away people's farm tools, women's cooking utensils, radios, bicycles, etc. . . . One day a young man in No. 5 was beaten for making charcoal; people couldn't take it anymore, so they rioted. And then the mgambo threatened to shoot the villagers. That's why the district police commissioner had to come and intervene."[44]

FIGURE 4.3. Public meeting with the *mgambo* (civilian paramilitary forces), the district police commissioner, and residents of Makaani, No. 4, and No. 5. Photo by Selemani, December 2015.

Despite the harm people suffered, few victims reported their cases to the police or sought medical treatment. For victims of assault, the only way for them to get free medical assistance from public hospitals was to file and obtain a police report, known as PF3 (police form no. 3); otherwise they would have to go to private hospitals and incur high fees, which they could not afford. More important than these financial and bureaucratic obstacles was people's lack of confidence in the state's ability to deliver justice. Many people were distrustful of the existing alliance between EcoEnergy representatives and district political leaders and knew that their odds of receiving due process were slim. Consider the following excerpt from an interview with a married couple, Zuberi and Beatrice, in which they raise this concern:

> ZUBERI: One time, [when I confronted the mgambo] they forced me to do those bunny hops, the kind of drills they make you do in military training. I refused. There were eighteen to twenty of them. There were many, I was just one. Ali was there, too, and he started commanding the mgambo to beat me.
>
> BEATRICE: We told Ali, "Farming is our only livelihood; we need land to live! We do not have money to live!" Then Ali smirked and said, "Why didn't you tell me so? If you need money, I will give you money so you can leave."
>
> ZUBERI: He told us to take the money and go. I don't remember exactly, but he threw something like TZS 2,000 or 5,000 [one to two US dollars]. I said I don't need that sort of help. It was very demeaning. . . . Ali, he is a small person with a lot of pride. He told us, "People like you who live in the bushes are like trash." He threatened to take us to the police. When you know that the mgambo is sent by the district commissioner and EcoEnergy, there is no point in reporting anything.[45]

When I broached the subject of mgambo violence with Ali and Saidi, they reacted defensively. Saidi, the more senior and outspoken of the two, stated that reports of beating were untrue: "I have not even seen a single police report. People can sue the police if they think they have been wronged. If they want to prove that they have been beaten, then they have to show that they have received the PF3."[46] His position was that if there was no formal evidence of assault, it did not happen, or at least that there was no way to prove it had happened.[47] However, to invalidate or ignore the "evidence of experience" would be to discount the far-reaching and irreparable consequences of violence.[48] Daudi's case below offers a sobering example.

Daudi

Daudi was a forty-five-year-old man who had migrated to Bagamoyo from Dodoma in early 2011 with his wife and two children. They lived together for a few years in No. 5, but he later sent his family back to Dodoma to live with his relatives for reasons of safety and security. In 2015, he began cohabiting with the woman he called his second wife, a recently divorced young single mother from Matipwili. I was first introduced to Daudi by Shabani, the chairman of No. 5. The three of us met on March 18, 2016. It was two days after the mgambo and Ali had assaulted Daudi. From the way he spoke and carried himself, it was evident that he was suffering from pain and struggling to process the trauma of violence:

> Whenever I try to explain it, I am overwhelmed by grief. . . . The day before yesterday I started to prepare my farm. When the mgambo saw me, they interrogated: "Who gave you permission to build this *banda* [farm shed]?" . . . There were five mgambo plus Ali. . . . They told me to lie down, facing the sun. It was around three p.m., the sun was still strong. I refused because I thought they could destroy me. That's when they started beating me around this part here [*rubs his right ribcage down toward the abdomen*], my arm, my eye. . . . They kicked me hard around the ribs. I don't think I am okay inside here [*cringes in pain*]. . . . I had never experienced such torture in my life.[49]

Shabani was a witness to this incident. He said he begged Ali to make the mgambo stop. Instead, Ali allegedly joined the mgambo and started beating Daudi in the knee with a baton. He remembered Ali telling Daudi, "Take this beating as medicine, *mzee* [old man], and stop making trouble."[50]

Daudi had neither reported his case to the police nor gone to see a doctor about his injuries. He couldn't afford the trip to Bagamoyo town or his medical expenses, but more importantly, he saw no hope for justice:

> There is no use in going to court. The police procedures and the government are no help to us. I remember there was another guy who was beaten. He went to the police, but nothing happened. I felt so bad, [I am] in so much pain, but there is nothing I can do. The mgambo and the police, they act like they are the owners of this country. Even our elected leaders, they act like they are the owners of this country. They came to us before and said: "Oh, we are working on this and that to improve your life," but they never came back and told us what exactly they have done. . . . We are only important to them during campaigns and elections; they need us when they need votes, but not now.[51]

Shabani also noted that people were reluctant to seek redress for fear of retaliation:

> When you go to the police, new and unknown cases will be fabricated against you, and as a result, you may have to serve time in jail for no reason. Sometime back in February [2016], several young men from No. 5 raided the mgambo campsite again out of anger. The mgambo retaliated; they continued to confiscate people's tools, destroy people's houses, uproot people's crops, burn people's charcoal furnaces, and beat them like they have done to Daudi. The mgambo captured several of the raiders and put them in jail. When they saw Daudi put up his banda, they accused that he, too, was an accomplice in the crime. Daudi didn't do anything wrong, and he didn't contribute to that raid, but he was scared to go to the police.[52]

On May 24, 2016, two months after the assault, Daudi passed away unexpectedly. He was reportedly rushed to the district hospital after experiencing pain and swelling in his abdomen. When his condition did not improve, he was transferred to the Muhimbili National Hospital in Dar es Salaam, where he died, according to doctors, from liver cancer. It is unclear whether there was a postmortem examination or if Daudi had indeed been afflicted with cancer all along without knowing. The news of Daudi's death spread quickly across the project site from Bozi to Matipwili, and everyone was gripped with grief and outrage. Those who had known Daudi were in shock and disbelief. He had been a healthy man just before the assault; to them, his death could only be explained by the violence he was subjected to at the hands of the mgambo and Ali. According to Shabani, a few weeks before Daudi's condition deteriorated, one of the new mgambo recruits who had not been involved in the assault but who felt bad about the situation came to apologize and to give some cash to Daudi so that he could seek medical treatment; when Ali found out about this, he allegedly fired that recruit.

I visited with Daudi's second wife six months after his death. She was still trying to make sense of what had happened:

> He was healthy before the assault. After the incident, he said it felt like his ribs were being compressed. He also complained of pain in his abdomen. He said he didn't know why he was beaten. I still don't know why. There is no one else to blame for his death. If EcoEnergy hadn't come to our area, this would have never happened. I am afraid, very afraid. I feel alone. After Daudi died, many people left because they thought the same thing might happen to them. There is no peace here, only fear and uncertainty.[53]

Causality is neither self-evident nor straightforward; it is always contested, and therefore always political. EcoEnergy denied being aware of any acts of violence such as those described in this chapter. I have endeavored here to honor the stories people shared, many of which were painful for them to relive and for me to reproduce. In closing, one is compelled to ask: How avoidable was Daudi's suffering and the suffering of others who were "kept alive but in a state of injury"?[54] In the interview two months before his passing, I had asked Daudi about his hopes for the future. His aspirations were humble and modest: "If my condition gets better, I shall rebuild my banda and return to preparing my farm. If we are allowed to expand, I would like to plant coconut and orange trees."[55]

In stark contrast to the deadlocked negotiations between the company, the state, and donors in Dar es Salaam, the project effectively turned people's lives upside-down in Bagamoyo. The management techniques the foreign consultant and district paramilitary deployed on the ground were qualitatively different and seemingly contradictory on the face of it. The skills trainings the former offered were meant to empower the "PAPs" to become self-reliant subjects, and the surveillance and intimidation tactics the latter used were intended to demonstrate EcoEnergy's exclusive rights to property. But similar to the PPC, these mechanisms of power both laid the foundation for the displacement of rural populations, whom the project considered an obstacle to completing the land deal.

The particular identities, social positions, and experiences of the consultant and the paramilitary informed how they understood and operationalized their respective biopolitical and necropolitical mandates. Their interventions, however, took on a life of their own as they came face to face with the contingencies of the project as well as existing cultural norms and power relations on the ground. As I demonstrated, they impinged directly on the lives of individuals, families, and communities in uneven gendered ways. This was evidenced through the gendering of livelihood options under the Early Measures, the conjugal conflicts and struggles over social reproduction that erupted as a result of women's participation in the training, the disproportionate paramilitary violence against men as the presumed household heads and decision makers on the land, and the masculinization of space with the presence of not only the mgambo but also the fathers who stayed behind to protect their families from harm, often at their life's expense. In the chapters that follow, I examine the highly differentiated forms of resistance that emerged in the midst of these changes, as people navigated the interstices of life and death.

5

NEGOTIATING LIMINALITY
Everyday Resistance and the Moral Economies of Difference

On September 21, 2016, Shukuru Kawambwa, a CCM politician and MP of Bagamoyo, visited Razaba at the request of a small group of young men who called themselves *kikundi cha wanaharakati wa Razaba*—the Razaba activists' group.[1] The visit was meant to be a listening session where the MP could hear directly from his constituents about their concerns over the increasing violence and insecurity in the area. The meeting was scheduled for 10 a.m.; villagers started arriving one by one, by foot, bicycle, and motorbike. By 11 a.m., there were roughly one hundred people gathered. People expected the meeting to begin an hour late as these things went, but even as the clock neared noon, there was no sign of the MP. Growing impatient, some people started leaving, many of them women with infants and toddlers in tow.

About a quarter past noon, a white SUV appeared on the horizon, churning a cloud of orange and gray dust in its wake. All eyes were on the MP as he got out of the car. He wore a tailored *kitenge* shirt whose vibrant colors and patterns stood in sharp contrast to the white car and the villagers' faded clothes. With a politician's smile, Kawambwa greeted the crowd: "Habari zenu?"—How are you all? Following a rehearsed script led by the young activists—the *wanaharakati*—the villagers chanted, "Fine, but not so fine. We are living here like animals! Like in the Congo! Like we are not here!" An old man from the back of the crowd interjected, taking everyone by surprise, "Waiting to be slaughtered tomorrow!" People's gaze darted toward the front again as Emmanuel, the wanaharakati spokesperson, shouted at the top of his voice: "Everything is broken. Our marriages, our legs, our arms, our houses, our hearts. Life is tough!" An old woman standing next to me mumbled under her breath, "Indeed, we are very troubled."

The smile on Kawambwa's face had faded by the time people finished sharing their views. To respond to an everyday greeting like "habari" in such visceral ways with an outpouring of frustration went against the culture of greetings in Swahili, where one was expected to be respectful and hold back negative feelings regardless of the circumstances.[2] Kawambwa empathized with the crowd, saying that he, too, had been born and raised in a village in Bagamoyo, and that he understood what it meant to live on the land. He asked what changes people wanted to see. Rama, the wanaharakati secretary, stood up and listed three key demands: an end to mgambo violence; land sovereignty (*uhuru wa ardhi*); and community development. He went on to elaborate:

> Our life has gone backwards because of EcoEnergy. The company has been using shady methods like bringing the mgambo to threaten us. The mgambo have stopped us from building and improving our homes. They have beat us and harmed us. Some people were thrown in jail for trying to survive. But we are still waiting for compensation. We cannot plan for the future like this. Our only livelihood is farming, but we have trouble farming, and we need land for farming. Our children are not being schooled. We don't know if our subvillage will be erased. We don't have a case against EcoEnergy, but we need answers. Honorable MP, why is our government turning a blind eye on rural citizens?

The air was thick with tension as Rama wrapped up his speech. People fidgeted on their feet, whispered inaudible words, exchanged impatient looks, and shook their heads either in disbelief of the situation Rama was describing or in disagreement with what he was saying. The old woman standing next to me grumbled again, perplexed, "But we *do* have a case against EcoEnergy, don't we?" Kawambwa offered no satisfactory response to Rama's closing question, only to remind people that he was just one of many MPs in the nation, and he did not have all the answers. The meeting soon adjourned, and a murmur arose from the crowd as people scattered, muttering the familiar refrain: serikali ni serikali.[3]

Raw feelings of abandonment, exasperation, and indignation permeated the words people spoke at the event. It was the most dramatic show of public outrage I had witnessed during my time in Bagamoyo. Those gathered seemed to have reached a boiling point after years of being told they don't belong, of waiting interminably for displacement, and of coping in vain with the injustices the project-affiliated mgambo perpetrated. They protested the land deal not as a single event but as protracted, messy, and confused processes and relationships that, literally and figuratively, broke people's lives and made them feel like animals awaiting slaughter.[4]

This kind of social mobilization the wanaharakati organized certainly had political power, by putting direct pressure on elected officials who were accountable to the people. However, although they pledged to work in the interests of all the residents whom the EcoEnergy land deal impacted, the wanaharakati, in their words and actions, reflected, and were constrained by, their particular social positions, spatial locations, political interests, and lived experiences as young educated migrant men, members of the opposition party, and residents of relatively new settlements within Razaba. As I later elaborate, the group routinely excluded from their activities and membership women, elders, and longtime residents, especially those with deeper historical knowledge and memories of the landscape, including traumas of prior enclosures discussed in chapter 2. Understanding the role of charismatic actors like the wanaharakati in contentious politics is important, but privileging their work risks devaluing the more subtle, submerged, and mundane ways in which ordinary people, as heterogeneous and divided as they were, translated their dissatisfaction into diverse expressions of agency.[5]

But conversely, the risk of interpretive analysis is that we might stretch, romanticize, or read too much into what might otherwise be ordinary behaviors and render them necessarily conscious.[6] Much of what scholars categorize as resistance—including what I described in chapter 2 as "everyday acts of presence"—is comparable to adaptation. James Scott called it "nibbling away" at power.[7] Resistance, in other words, is akin to the normal stuff of everyday life, where ordinary people, as individuals or in groups, try their best to cope with, make sense of, and complain about oppressive situations and structures, while improvising and strategizing ways to secure their survival and recognition. Motivations behind resistance are often difficult to pinpoint, because the reasons people do what they do are rarely singular, static, or straightforward; intentions are better understood as plural, evolving, highly subjective, and at times contradictory.[8]

Attending to these complexities and nuances, this chapter examines what I have identified as three key repertoires of resistance in Bagamoyo: ordinary speech acts, illicit practices, and political organizing. Though ordinary speech acts comprise an expansive field that includes muted grumblings and discontented murmurs like those of the elderly woman described in the opening vignette, I focus here on rumor and gossip as the most common and gendered means through which people engaged in moral debates and collective sensemaking. Moving from the discursive to the material realm, I then go on to examine three illicit practices people adopted in response to the paramilitary repression of agricultural livelihoods: clandestine farming, charcoal making, and alcohol brewing. Those who engaged in these activities were some of the poorest of the poor, such as elderly widows, divorcées, single mothers, and migrant youth; as I noted in chapter 4, many of them resided in areas where the mgambo maintained the heaviest hand and thus were disproportionately subject to arbitrary punishment. Despite the mgambo's threats and criminalizing of their livelihood practices, they pressed on regardless, defending their actions as necessary and justifiable for their survival. Finally, in the last part of the chapter, I return to the political organizing by the exclusively male wanaharakati, investigating the group's origin, membership, and mobilization strategy that hinged centrally on claims to "agrarian citizenship." I also analyze their internal fragmentation, as well as tensions with other local residents, which arguably limited the longevity and progressive potential of their activism. These three repertoires of resistance are not mutually exclusive, but I discuss them one by one to illustrate their particularities and the politics of difference on which they rest.

Ordinary Speech Acts

In her study of vampire stories people told to describe colonial power in East Africa, Luise White has argued for an expansion of historical epistemologies that include rumor and gossip "to find the very stuff of history, the categories and constructs with which people make their worlds and articulate and *debate* their understandings of those worlds."[9] If we consider ordinary people as historical subjects, then their everyday narrative practices, including rumors, gossips, jokes, and metaphors, merit consideration as historical sources in their own right. In Bagamoyo, people gathered and shared stories (*-piga*, story) in a variety of gendered spaces, including the bars, markets, and party meetings that men frequented, and the homes, kitchens, and gardens where women assumed primary responsibilities.[10] Spoken beyond the direct gaze of state and corporate actors, the conversations people exchanged in these spaces functioned as "hidden transcripts" or, in James Scott's terms, "nonhegemonic, contrapuntal, dissident, subversive discourse."[11] By articulating their thoughts and grievances into words, in however abstract and incomplete ways, people sought to gain a modicum of autonomy and control under conditions of so much uncertainty and upheaval. Rumor and gossip, in particular, became common vehicles through which men and women speculated about the near future and debated which information they considered true or false and which events and possibilities they thought were just or unjust. These otherwise prosaic speech acts afforded people an opportunity to seek out "truths" by assembling pieces of "raw, confused facts" and lay out theories about what could, what might, and what ought to happen next.[12] The stories people shared, as I show, always had a moral and gendered edge to them.

Rumor

Rumor is not necessarily false news, but more in the nature of uncertain or unverified information.[13] When the Early Measures came to an abrupt halt, and when the foreign consultant and her team stopped their site visits around mid-2015, people suddenly found themselves cut off from formal channels of communication with the project (see chapter 4). Though the mgambo still remained and functioned as agents of EcoEnergy and the state, people were reluctant to take their words seriously. As one elder put it, "anayekuja pasi na hodi, huondoka pasi na kuaga"—a person who comes without a knock or whose manners are untransparent is not to be trusted.[14] A near absence of reliable information meant that people's standards of what was plausible began to shift; they became more suggestible and receptive to rumors than they would have been otherwise.[15]

The types of rumor that had the widest circulation were those that offered people a temporary reprieve from the restlessness of waiting—waiting for compensation, displacement, and resettlement. At any given time, there were multiple rumors in circulation. This was due in part to the way rumors spread. With their repeated transmission and dissemination over time through complex chains of anonymity across different locations, rumors were open to interpretation, distortion, omission, improvisation, and elaboration with whatever new information became available. The result was a coexistence of not only multiple but also conflicting rumors with incongruent temporal projections.

For example, between 2015 and 2016, a rumor circulated that compensation and displacement would commence concurrently in April 2016. Another rumor suggested that although compensation would be indeed paid in April 2016, actual displacement would not occur until two months later. Yet another competing rumor said no one would be evicted until everyone had finished harvesting their crops in July and August 2016. Of all the rumors, the last version seemed to gain the most traction; as much as people desired to remain on the land, they considered it to be the most plausible and also the fairest outcome. Consider the following conversation that took place at a bar in the settlement of No. 4 in late March 2016, between the bar owner Mohammedi and his patrons:

> MOHAMMEDI: The mgambo are using force to drive us out of here. Their strategy is threat and intimidation. Otherwise, how would the project move us next month?

MAN 1: That's not true. I remember the white woman [Anna, the foreign consultant] saying one time that we won't be forced out until after the harvests. We just finished planting. Are we to just abandon our farms?

MAN 2: Who knows where we'll be moved to? *Lo!* Who knows we'll be given land? Better we be evicted with some food to eat.

MAN 3: Exactly. If they don't even let us do that, what's the difference between us and refugees?[16]

Plausibility is a matter of degree and a function of cultural context.[17] In debating different rumors, the men at the bar were making normative assessments about what they perceived to be right or wrong, appropriate or inappropriate, or true or false, based on prior knowledge, experience, and existing norms of subsistence. That is, they were making moral arguments about how displacement, however undesirable, ought to happen in ways that ensured the rights of subsistence producers, for whom access to land and food were vital. To be evicted hungry, homeless, and landless would be a bitter blow to their dignity and identity as farmers and citizens. In the postcolonial "national order of things," the category of "the citizen" signified the most pure and prestigious form of belonging, and "the refugee," the most polluted and unworthy.[18] It would be wrong to be disgraced to a refugee-like status on their home soil.

All rumors surrounding the temporality of displacement, however, fell apart when the news of the government suspending the EcoEnergy project, for the sake of wildlife conservation, began to spread in May 2016.[19] But as the mgambo violence went unabated and as villagers received no official confirmation from district authorities about the government's decision, new rumors quickly began to swell in the streets—rumors that EcoEnergy and the national parks authority, TANAPA, were one and the same; that it was TANAPA who had brought in the mgambo; and that TANAPA would evict people without compensation.

The effect of rumor in Bagamoyo was arguably twofold. On the one hand, rumors allowed certain groups, namely men with access to nominally public forums, to create and disseminate knowledge in ways that affirmed their collective moral values and allay fears about arbitrary displacement. On the other hand, the unstable nature of rumors magnified existing anxieties and insecurities. As one man told me in an interview, frustrated with the spate of rumors that brought him no relief, "Rumors are just rumors [*tetesi ni tetesi tu*]. We are still waiting for real answers."[20]

Gossip

Gossip is a broad narrative genre in which a few close acquaintances talk casually about an absent third party in ways that can be malicious, friendly, funny, or simply informative.[21] By reporting on specific behaviors and events, and evaluating the reputation of the absent entity, gossip can solidify group bonds and boundaries, and, like rumor, reinforce shared moral values.[22]

Nearly everyone I knew in Bagamoyo gossiped, but the spaces in which they gossiped and the topics they discussed varied. Much of the earlier cited conversation that took place at Mohammedi's bar qualifies as gossip. When Mohammedi and his friends invited me to come chat with them, I eased myself into the group by asking them where the women were. They chuckled and said that women were at home taking care of children and housework. One of them more bluntly stated that women were not considered trustworthy to partake in important conversations unless they were schooled. Though men clearly gossiped among themselves about various things, they associated gossip about private matters or secrets of domestic life as something women did.[23] They dismissed such talk as trivial and morally problematic, similar to the way gossip, in contemporary English use, is often gendered as "women's talk," something trifling, loose—something one should avoid.

It is worth mentioning here briefly that the negative connotation of gossip, however, is a relatively recent phenomenon; no references to gossip as idle talk, for instance, appeared in the Oxford English Dictionary until the early nineteenth century. Gossip, from the Old English phrase "God sib," originally referred to godparents and god-siblings, and later took on the meaning of female friends and relatives who supported one another in childbirth.[24] In birthing chambers and other feminized spaces of the home, gossips (that is, women's gossip circles) enjoyed a degree of autonomy from male surveillance and helped forge important social alliances. With the growing male anxiety over women's social power throughout the sixteenth and seventeenth centuries in England, gossips were defamed as liars and drunks, and "gossiping" was denounced as an unproductive social activity.[25] Using gossip as a category of analysis thus requires recognition of its patriarchal historical construction and an intentional attempt to undo its contemporary sexist stereotypes.

At the bar, the men talked at length about the mgambo. "EcoEnergy people are not human [*sio binadamu*]. They are our enemies [*maadui*]," said Mohammedi in reference to the mgambo assault on Daudi that had happened a week prior (chapter 4). Mohammedi was not the only one who attributed Daudi's death to mgambo violence and, by extension, the EcoEnergy project.

"Did you hear that they burned down Daudi's shed before they beat him?" asked the man sitting next to Mohammedi, incredulous.

"They are savages [*washenzi*]," replied one man.

"They are liars [*waongo*]," said another. "One day they tell us they are bringing us jobs and development, the next day they bring us chaos and death." The men's assertions pivoted on an us/them distinction.[26]

This kind of name-calling, backbiting, and general malicious talk comprised a popular subgenre of gossip. "Troublemakers" (*wakorofi*), "tricksters" (*wajanja*), "fools" (*wajinga*), "thieves" (*wezi*), and simply "bad people" (*watu wabaya*) were other words people used frequently to disparage EcoEnergy, though it was not always clear what specific actors they were referring to. Rhetorical devices, such as "Did you hear . . . ?" or "Don't you know . . . ?" were also common ways through which people expressed their own moral stance and gauged one another's.[27]

Beyond malicious talk, humor was another subgenre of gossip. Through telling funny stories, people ridiculed and critiqued the way EcoEnergy attempted to build relationships with local residents through various community events. Consider the following story narrated by Omari, a young male farmer and *pikipiki* (motorbike) driver from Bozi. He was resting under a tree near an EcoEnergy project sign with several other drivers, none of whom were from the area. Omari and I were supposed to meet for an interview that afternoon. When I arrived he was midconversation, telling his fellows about a sporting event EcoEnergy had organized a few years earlier. He gestured me to come sit with them as he continued:

> I waited and waited. Years passed, but nothing happened. Then out of the blue in 2014, EcoEnergy came and told us to come play football! They took young men from everywhere from the project area. We were surprised. We were given jerseys that read "EcoEnergy Bagamoyo." They took us to a field in Kiwangwa [a village outside the project area, which the company had targeted for its outgrower scheme]. The grass was bad, and there was sand everywhere. I didn't have proper socks and shoes. And it was two p.m. Playing football, barefoot in the sand, at two p.m. Imagine! It was *hot* and *painful*! I couldn't play easily. I told them, "Hey, I cannot play anymore. Give me some socks at least." They gave me socks, but they didn't really help, so I had to say, "Hey, please stop, stop, stop." Then I was substituted for good. Some people had socks and shoes, so they could play for longer. At the event, they took a lot of photographs. . . . My team lost the match. But you see, we were only recruited just one day before the match, and we weren't given any time to practice beforehand. They just rounded up some young men and

said, "Don't let us down." They promised to feed us, and so we went, but *lo*! There were only three small pots of *bokoboko* [soft porridge]! [*Big laugh*][28]

I heard a few different iterations of this story from other young men, including the members of the wanaharakati, all told in hindsight with an infusion of humor. In thinking that the event was funny, those who narrated the story were not simply making light of a lousy situation. They were passing moral judgment on things like how the players were recruited at the last minute; how they were given jerseys but no other equipment; how they were forced to pose for the photographer wearing the branded jerseys; and how there was not enough food to eat. Recounting his experience participating in this "Football Bonanza," as EcoEnergy called it on its corporate website, Emmanuel laughed and said, "EcoEnergy has a habit of cooking things up. It's been their style. They do things to catch people's attention."[29]

Women, on the other hand, gossiped in more private settings, in pairs and small groups with their female neighbors, friends, and relatives. They talked while they cooked meals, picked vegetables, nursed, and fed children, collected water and fuelwood, did laundry, and wove mats and baskets. They would share with one another the latest news and rumors they heard from their husbands or local leaders; and like their male counterparts they engaged in debates and speculation about which rumor was "real" or which seemed the most plausible. After the women exchanged ritualistic greetings and shared news, their conversations would often flow into more personal subjects, concerning their marriages, families, and communities. The topics women discussed were not strictly domestic matters, as men assumed, insofar as they were also intimately tied to extralocal forces that shaped their lives.

For example, following the PPC, I observed married women share with one another their dual fears of losing land to EcoEnergy and losing compensation to their husbands. Other times, they complained of how their husbands were spending too much time socializing with their friends, or how their husbands said they were going to town to "get news" only to return late, drunk, with no news, ultimately increasing the workload for women at home and at the farm. Women also confided in one another about their husbands' extramarital affairs, which they often attributed to the instability and restlessness created by the project. Mothers and older women raised concerns about the growing number of men migrating to the area in recent years and the implications this had for community gender relations and the safety of their daughters. They at once condemned and empathized with young women who they thought were selling their bodies to migrant men, or those who ran away with them to escape the binds of poverty and precarity. Women's gossip, therefore, was not some insignificant tittle-tattle. Gossip was an important means through which women grappled with all the complex ways the stalled land deal impinged on their daily conjugal, family, and community relations. The moral economy of gender relations was at the heart of women's gossip and grievance about the project.

Illicit Practices

Fleeing or evading forced labor, withholding tax payments, and occupying so-called state-owned resources have been some of the recurrent motifs in the literature on peasant politics.[30] Though these were means by which peasants sought to safeguard their lives and livelihoods, the state always held the prerogative of labeling and prosecuting such acts as immoral and criminal. As I have repeated throughout, although people living within the EcoEnergy concession claimed the land they cleared, occupied, and used as rightfully theirs, state and corporate actors frequently cast them as intruders or mere squatters on public land. "The correct term for those people is 'intruders'" (*wavamizi*), as a senior Ministry of Lands official corrected me during an interview when I referred to them as "residents" (*wakazi*).[31] Yet, in the eyes of local people, compulsory land acquisition, livelihood restrictions, and mgambo violence were equally criminal, as they violated what people perceived as their right to subsist. Criminality, hence, was as much about perception as it was about the law.

Regardless of their perceived land rights, as the social conditions of farming became increasingly precarious over time, some of the poorest residents most affected by paramilitary activity resorted to illicit livelihood practices. They improvised clandestine or guerrilla-style farming strategies and took on other unauthorized resource-based livelihoods, such as charcoal making and alcohol brewing. Those who engaged in these works were well aware of the potential consequences of being caught by the mgambo, but they carried on anyway because they saw no alternative, and they believed their actions were morally defensible.

Clandestine Farming

"Farming in secret" (*kulima kwa kuficha*), as people called it, entailed clearing the land and sowing at night, or during times when the mgambo seemed inactive or less active.[32] To make this strategy work, people first had to identify patterns of mgambo surveillance. But this task was not as easy as many imagined. The times the mgambo patrolled and the routes they took were often irregular. People had a rough sense of when the mgambo might show up, but estimations varied, even among those who lived on the same street. "They come here every day. They come around nine or ten a.m. and then again around three p.m.," said one woman in No. 5.[33] Her neighbor next door, on the other hand, said, "It depends. Sometimes they come every day, sometimes they are quiet for several days and then show up out of the blue. One day they might come in the morning, another day they might come in the afternoon or early evening. It's confusing. Sometimes they might stay in their camps but not actually do any patrolling."[34] Her husband added, "Sometimes you might find them waiting for you at your farm ready to confiscate your hoe."[35]

Apart from their inconsistent patrol schedules, other difficulties arose: there was a high turnover rate among the mgambo, and they often rotated from one campsite to another, making it difficult for people to build relationships with any one mgambo and to "understand their habits" (-*kuelewa tabia zao*) well enough to predict their next move.[36] Surveillance remained effective because of its impersonal nature. Yet there was another obvious problem: farming could never be kept underground. "The land will show. The land cannot lie. Doesn't matter if you succeed in planting at two a.m.," said one woman in No. 5.[37] Many people shared her view, describing how they were playing a game against nature they could not win. As one man related, "The land does not lie. Come rain or shine, crops will start to grow. The mgambo will see the crops and ask, 'Who gave you permission to farm? You are not following the orders!'"[38]

When confronted by the mgambo, most people feigned ignorance and innocence, while bracing for retribution. To circumvent confrontation with the mgambo, some people resorted to clearing new fields in the forest far away from their homes and existing fields—although, even in this case, people still risked losing their crops if the mgambo later discovered them. When all else failed, some decided to leave behind their fields *and* homes temporarily so as to give an impression of voluntary abandonment. These included people like Aziza and Musa, a couple in their mid-forties, who first migrated to Bagamoyo in 2010 from the Tabora region in western Tanzania. I first met them in No. 5 in 2014, but the next time I saw them in 2015, it was over on the hillside, opposite the western border of the project area along the railway tracks. Musa explained why they had shifted:

> When the mgambo came with guns and started torturing people, everything changed. We initially had two acres, but we wanted to increase to three acres so we could also grow watermelon for sale. We wanted to plant papaya, too. The reason we left—there is no other reason than the mgambo. They told us not to expand our farms. Then they said

we shouldn't be growing food at all; we shouldn't build; we shouldn't repair our homes; we shouldn't do this or that. We felt constricted, we felt chased.[39]

Over on the hillside, the couple set up a makeshift hut and cleared an unoccupied area of land, about an acre or even less, just enough to grow maize for food. Though they may have been successful in escaping the mgambo's gaze, their livelihood and living conditions were no less constrained than before. The land was less fertile, and they were farther away from a water source. With a look a shame in her eyes, Aziza described how they had been eating unripe green maize and wild greens to curb hunger that year. The couple saw themselves as farmers above all else; constantly being chased from the land and not being able to feed themselves felt humiliating. They recognized, however, that the shame was not theirs alone; the government had to share in it. As Aziza made clear,

> The problem here is this: We are being ruled. Democracy is on paper only. We have rights, we have been counted [during the PPC], but the government rules those rights with force. Small farmers have been forgotten and are called intruders, enemies, criminals. . . . In Tanzania, farmers should have access to land. We should be able to live anywhere, as long as we are not breaking any laws and working the land with our hearts. We are farming maize, cashews, and mangoes, not selling *bangi* [marijuana] or *gongo* [strong distilled liquor akin to moonshine]. The land cannot lie. The land tells you who its caretakers are, who its workers are. What has the government done for the land but to give it to an investor who, so far, has done nothing? Who, if not the state, is the criminal?[40]

In her view, working the land was the most earnest and honest thing one could do. Instead of redistributing the land to hardworking rural people who truly need it, the government signed away the land to an investor who had nothing to show for it.

Charcoal Making

Charcoal making was occasional seasonal work for longtime residents, but for many migrants it was sometimes their only means of livelihood after they nearly gave up farming. Yet only those who obtained permits from the district forestry office could legally produce, transport, and sell charcoal. This existing regulation provided the necessary legal grounds on which EcoEnergy prohibited charcoal production within the concession area. As Saidi, the company employee explained,

> People are not allowed to cut down trees in the project area. And by law, you are not allowed to cut down trees for charcoal unless you have a permit from the district commissioner. Whether it is for business or domestic consumption, you need a permit. So first you need permission from us to use the trees in the project area, and then you need a permit from the district to cut the trees for charcoal. The district has not issued any new charcoal permits since July 2015.[41]

Most charcoal producers were young migrant men, although a few women engaged in the practice, too, under conditions of extreme economic hardship. First-time women charcoal makers typically worked alongside the men to learn the ropes of the trade, from preparing the kiln to bagging the charcoal, in return for a small share of the proceeds. While a common perception across Africa is that charcoal production provides quick cash for the rural poor, many workers contested this notion. As one man explained during a focus group discussion, "Charcoal [making] is not an easy job. We don't love it. We are working with fire. It is dangerous. We get TZS 16,000–17,000 [approximately eight US dollars] per hundred-kilogram bag [220 pounds]. It's not bad, but our health is suffering because of it. Our chests are in pain because of the dust, smoke, and gas we inhale."[42] For novice female producers like Rukia, a divorced mother of three (introduced in chapter 3) who had described charcoal making as "fever-inducing work," health was also a major concern. But this concern was compounded by other constraints, such as gender-based exploitation by traders and middlemen, as well as the responsibilities of child care as a single parent:

> This work is seriously affecting my health. When you burn and harvest charcoal, you are directly playing with fire. I don't wear any protective gear, so you go into your furnace, and you feel the heat through your entire body. It's difficult to breathe. I hear that some charcoal producers drink milk afterward to wash down the ashes, but I don't have that kind of money. When I am done harvesting and come home, I feel like I have a fever, like I am ill. I can't go anywhere; I can't do any work. Sometimes my eldest daughter helps out, but that means she also has to bring along her younger siblings and sit near the fire, which is very dangerous.
>
> Some people are doing proper business with charcoal, but I am just doing it to survive. I sell to traders from Zanzibar. I call them when my charcoals are ready for harvest. They come see my furnace, check the quality of my products, and then take them if they like them. They

pay me between TZS 5,500 and 7,000 [roughly three US dollars] per hundred-kilogram bag, enough to buy me salt. But the whole process of making charcoal—felling trees, gathering all the materials to build a kiln, burning, unloading the charcoal, sorting, and bagging—it can take me up to two and a half months, because I do it alone. Young men can do it in two weeks, and some of them burn more than one kiln at a time. They also know more about the business, and they are less likely to be cheated by middlemen. I am not sure if what I am getting is a legitimate amount.[43]

Rukia was nervous about the mgambo apprehending her one day and what that would mean for her children. It was not a livelihood she was proud of or wanted to continue: "If I had the choice to stop doing this, I would. I would rather plant trees than to cut down trees."[44] Though she was never caught, other charcoal producers were less fortunate. They had their kilns destroyed and bags of charcoal seized; some were arrested and taken to prison. Others were able to get past these penalties by paying the mgambo bribes in advance. The male charcoal producer quoted above said, "If you give the mgambo some money, they will leave you alone. Some mgambo won't accept it, though. When you are taken to the police station, you can try to pay another bribe there if you have the means. If not, you'll be jailed. Some of our men are already in jail."[45]

According to Saidi, the district government was not so concerned with these unlicensed individual charcoal producers as it was with organized smuggling operations: "Charcoal is a *big* business. It's big enough of an issue to cause terrorism, like you hear about the illegal charcoal trade in Somalia. I can say that 90 percent of the charcoal that is made here in Bagamoyo is being illegally smuggled to Zanzibar."[46] None of the local charcoal producers I interviewed admitted to being part of an illicit trade network, but most people knew about the illegal port (*bandari bubu*) near Winde where the smuggling reportedly took place (figure 5.1).

"Why does EcoEnergy care so much about charcoal?" Mohammedi asked me one day as he took me on a tour of the port. It was a question many people asked, and one that some, Mohammedi included, answered only with suspicion:

EcoEnergy must be involved in some other activity. Why should they be interested in whether or not charcoal is being smuggled here? It should be the Tanzania Revenue Authority that should be worried about smuggling, not EcoEnergy. They are working with the district commissioner and looking for opportunities to tax people and make side money while causing us trouble.... Why would EcoEnergy and the mgambo act as if they were street police, chasing after charcoal makers and harassing our women for making *pombe* [beer] and gongo?[47]

FIGURE 5.1. Boats lie at anchor at *bandari bubu* (an illegal port) in a tidal channel near Winde. Charcoal bags are stacked on the bank. Photo by the author, September 2016.

Alcohol Brewing

Alcohol production was largely if not entirely women's work. It was typically widows who brewed and sold *wanzuki* (honey beer) and *komoni* (maize beer), in addition to other commercial alcoholic beverages and prepared foods in their home compounds. Younger migrant women, on the other hand, made and sold gongo mostly when their financial needs were dire; they would put the liquor in recycled water bottles or commercial alcohol bottles (e.g., Chibuku, Kiroba Original, or Konyagi bottles) and distribute them underground to individuals or local bar owners. Compared to other traditional brews, gongo is more toxic, and people and authorities alike see it as a major problem in both rural and urban areas, especially among the male youth.[48] Recall how Aziza considered selling gongo as an undesirable and dishonorable act compared to farming.

As widely documented, production of alcoholic beverages for sale has been an important source of independent income for women in Africa across time and space. Beginning in the early colonial period, women relied on beer brewing, food vending, and prostitution in urban areas, taking advantage of market opportunities created by the male migrant labor system. Women who worked for independent income, and thus transgressed the boundaries of normative gender relations enforced by Victorian ideals of domesticity, were often subject to police harassment and labeled as immoral by colonial authorities and white settlers.[49] In the postcolonial period, deteriorating economic conditions and declining producer prices for cash crops forced many rural households to diversify their livelihood strategies, which included, for many women, making careful decisions to divert food sources into alcohol production for higher returns.[50]

Neema was one of the brewers I met in Bozi. She was a widow in her fifties, who lived with her disabled mother. Neema had been selling wanzuki and komoni since the 1970s. Her late husband was a livestock handler at RAZABA, and most of her early customers were his fellow ranch workers. As Bozi grew over time, her customer base broadened to include local farmers, fisherfolk, charcoal makers, and other rural workers. She sold both food and beverages, and this offered her a steady source of dry season income for many years until recently, when business began to decline following EcoEnergy's arrival (figures 5.2 and 5.3). As she explained,

> I have been having fewer and fewer customers since 2011. And in 2014, the mgambo put up a gate [about one hundred meters from her house].

FIGURE 5.2. Alcoholic beverages for sale at Neema's, including Chibuku, Kiroba, and home-brewed honey beer inside the plastic bucket. Photo by Neema, August 2016.

FIGURE 5.3. Fried prawns and catfish for sale. Photo by Neema, August 2016.

> There was a lot more foot traffic here in the past, and many workers in the area were my customers. The mgambo put up the gate to stop new people from coming in and to restrict the charcoal trade. The charcoal makers went somewhere else, so I lost my customers.[51]

Neema's physical proximity to the mgambo made it difficult for her to go about her everyday activities. Making home brews required various raw materials, including grains, fruits, tea leaves, sugar or honey, yeast, water, and, and importantly, cooking energy. As she highlighted in an earlier photo-narrative (figure 0.4), she felt especially nervous about collecting fuelwood. It was not uncommon for the mgambo to harass women and girls who were on their return walks home, and the general precaution was to be fast, discreet, and always go in groups. "The mgambo are shaking things up here and creating a chaos," she sighed.[52]

In early 2016, for instance, the mgambo rounded up two women in No. 5, Mariamu and Tatu, for making and selling alcohol. Mariamu, a single migrant woman in her twenties, was reportedly caught selling marijuana and a twenty-liter bucket of gongo, and was imprisoned as a result. Tatu, a married migrant woman in her thirties, on the other hand, was said to have been wrongfully arrested. I met her soon after she returned home from police detainment. Her husband, Shabani, the chairman of No. 5, first described what had happened:

> The day she got arrested, the mgambo were busy searching for this young lad who was supposedly causing them trouble. They searched all over No. 5, including ransacking our house and shop. They couldn't find the guy. Instead, they decided to take Tatu, saying that she was selling gongo. But in fact it was *them* who had brought the gongo! They made it look like it was inside our shop. I followed up with the police and told them there was no evidence, and that I would take the case to court. I knew one of the judges, so I pleaded with her, and she ordered that Tatu be released.[53]

Tatu confessed she had made gongo in the past when money was tight, but she had not done so in recent years. She knew how potent the substance was and regretted being associated with it. But she also knew that there were larger structural issues that made people, both consumers and producers, dependent on alcohol:

> Gongo is often made with sugar and rotten fruits like papaya. In the past I've made it with stale *ugali* [thick porridge, usually made with maize], coconut, cabbage leaves, and rotting banana. Instead of selling gongo, others might sell gongo-infused cashews. They look like normal cashews, but they can intoxicate people to the point of unconsciousness. You could get drunk badly or even die with only just a handful of nuts.... The problem of drunkenness is a serious one and has been getting worse here.... Men drink gongo, and women, too, because they are frustrated; and women make gongo because they are also frustrated. It's a vicious cycle.

"What are they most frustrated about?" I asked. "The mgambo, the uncertainty, the project," she promptly replied. As we talked, she pointed toward a drunk man in the distance, who rose to his feet and staggered until he reached a tree to prop himself up.[54]

Political Organizing

The wanaharakati comprised a group of six young men between the ages of twenty-six and forty-one. Four were single, although one of the married men now lived alone after having sent his wife and children away when the mgambo violence intensified. Whereas most residents had only attended some primary school, all the members of the wanaharakati had completed secondary education. All were also relatively recent migrants to Bagamoyo. They arrived sometime around 2010 after having worked for wages at different jobs and locations, for example as a schoolteacher in Kigoma, a security guard at a diamond mine in Shinyanga, and a field assistant for an NGO in Dar es Salaam. Like most other migrants, they hoped to make a good living out of farming in Bagamoyo.

The men formed the group in March 2016 in response to the mgambo offensive and the district authorities' recent suppression of civil and political rights. A month earlier, in February, the district commissioner had placed a moratorium on local elections in Razaba, saying that the subvillage was in the process of being eliminated. Samwel, the chairman of the group, invited me to meet the activists one afternoon in July 2016. When I asked him at the meeting how the group came about, he explained, "We created this group to stop the oppression and corruption of EcoEnergy and the district commissioner. Our goal is to reach the central government. We don't trust the district authorities." Borrowing the then-president Magufuli's anticorruption philosophy, *kutumbua majipu* (bursting the boils), Samwel argued,

> There are so many *majipu* in the district. We want to meet with policy makers at the national level, even the prime minister or the president. When EcoEnergy found us here, they promised they were going to make our lives better. But their promises have not been fulfilled. People are unhappy because of the beatings by the mgambo, the prohibition on farming, restrictions on building our homes, and the stalled local election.

Rama, the group's secretary, echoed Samwel's comments:

> The government wears spectacles made of wood—they cannot see anything. They are just pretending to see the citizens. Our leaders think that citizens here are to be colonized, to be oppressed. They are surprised that we even exist. They think we are bandits, drunkards, and drug addicts. All we want is to get our rights back. We want our subvillage to be restored. We are about working together for the common benefit of all villagers.

When asked what specific rights the wanaharakati were fighting for, Rama replied, "The right to live in peace. Let us live in our subvillage as we should." Majid, the deputy spokesperson, chimed in: "We should be recognized as humans. We are living here not against the government's plans but as legitimate citizens. They shouldn't discard us like trash." "The government should take EcoEnergy elsewhere and give the land back to the citizens," commented Abdallah, the group's treasurer.[55] The wanaharakati drew on moral claims to citizenship, particularly what they referred to as agrarian citizenship, hoping it would serve as a rallying call for everyone on the land whom the project aggrieved. Yet, as I show, their chosen idiom of collective identity would only go so far in terms of uniting people across difference.

Peasant Intellectuals

Each member of the wanaharakati brought to the group his own experiences of confrontation with the project. Samwel was once caught and arrested for making charcoal; he had to pay TZS 50,000 in fines and serve three months of community service. The mgambo confiscated Abdallah's farm tools as a result of a quarrel he had with a local EcoEnergy staffer. Another group member, John, received multiple threats of incarceration for expanding the area he farmed; he reported his complaints to the district militia adviser, which only made him even more exposed and vulnerable to surveillance.

Emmanuel's encounter with the project was quite different from those of other members. He had previously worked at EcoEnergy's seed cane nursery, which the company set up on a two-hundred-hectare parcel leased from a prison near Bagamoyo town. He started out as a storekeeper and was later promoted to an assistant security officer. He said that during the two years he worked there, he tried to advocate for higher wages for the casual farm laborers. Between 2012 and 2014, he said the cane cutters were making TZS 3,500 per day, while the figure I received from EcoEnergy was TZS 4,900; either way, the daily wage was negligible, around two dollars or less. In a separate interview, Emmanuel explained at length his concerns about the labor conditions at the nursery, and how he and other workers eventually decided to clear small plots on the nursery periphery to grow their own food:

> The white farm manager was oppressing people so much. He was mistreating people. I was earning TZS 250,000 [USD 125] per month, given my education level, but it was not enough to support my wife and child, who were living in Bagamoyo town at the time. However, I was more shocked at how little others were getting paid. I thought they probably could earn more on their own farms. They were fathers and had families to support elsewhere, like I did. I saw that the more they worked, the more they suffered economically. White people owned the company [EcoEnergy], but they had another company to recruit laborers. When I raised my concern to the recruiters and the farm manager, they said it was none of my business. They also said they were providing

the workers with morning porridge and lunch. But you see, these meals were not free. They were deducted from the workers' wages!

Eventually the workers, including myself, decided to clear small patches of land near the nursery so that we could at least cultivate our own food. But later, the manager, the mgambo, and the prison police chased us away. We were told that we were invading the farm. As their leader, I was arrested. But I pleaded with the prison officer, and he gave me six months to harvest and leave. But before the six months expired, we were evicted from our plots. I did not want to go back to Shinyanga.... So I came here to Razaba to find an area to farm, without realizing that EcoEnergy had a claim over this land. Starting anew here was difficult at first, so I worked for about four months making charcoal and helping people at the bandari bubu [illegal port] until I had some capital to start clearing the land and build a house. But after a while, I started noticing the same things that happened at the nursery. The mgambo were doing the exact same things here, oppressing people, but at a much larger scale. That's why I decided to join hands with fellows like Samwel. We said, "Let's begin our work immediately tomorrow so we can reclaim our rights as citizens."[56]

The wanaharakati tried a number of strategies, including writing a complaint letter to the regional commissioner, deliberately bypassing the office of the district commissioner, which likely enraged him. Each member chipped in some money to hire a lawyer in town to help draft the letter. The outcome was not favorable, however, as the regional commissioner delegated the issue to the district commissioner, exacerbating already existing animosities between district authorities and the wanaharakati. Beyond trying to engage directly with state officials, the group also reached out to a local public radio station in Bagamoyo to air their concerns and raise public awareness about what was happening in Razaba. They also contacted a journalist, who covered their story on a nationally televised news program. At the grassroots level, they started touring different communities to make themselves known to their fellow citizens, listen to their concerns, educate them about their rights, and mobilize their support. The wanaharakati, in this sense, acted as what historian Steven Feierman, drawing on Gramsci, called "peasant intellectuals."[57]

In their meetings and listening tours, the wanaharakati frequently invoked the notion of the agrarian citizen, or what they variably referred to as "rural/village citizens" (*wananchi wa vijiji*), "farming citizens" (*wananchi wa kulima*), or "citizens of the countryside" (*wananchi wa porini/shambani*). They called for distributive justice or land sovereignty: the idea that rural Tanzanians, by virtue of their being born on Tanzanian soil and working that soil, ought to benefit from secure access to land. They were not making nor could they make any claims or appeals to their deep connections to the land and people of Bagamoyo, given their migrant status. Rather, they upheld the idea of agrarian citizenship as a broadly shared mode of belonging to the land. These moral arguments resonated with many people, especially other migrants, but ambiguity and nervousness permeated the group's interactions with longtime residents and women, who questioned their representativeness and legitimacy.

Fault Lines

The longtime residents of Bozi were the most skeptical of the wanaharakati and their organizing. For instance, Athumani, the acting chairman of Razaba, was sympathetic to the group's efforts but unconvinced that they could succeed in any way without achieving a broader base of support: "The wanaharakati need more people in the front lines. They actually need people who have lived here for a long time if they want to be seen as legitimate. If they organize themselves like they are doing now, it will not be enough. They will lose. But in truth, I don't understand the significance of their activities. They have formed themselves without a full understanding of the history of the area here."[58] Another longtime Bozi resident expressed similar sentiments: "I told the wanaharakati, first, you need more people to fight with you. Second, you need to include both young and old people. And third, you need to work with people who have lived here for a long time. If they [the wanaharakati] organize the way they are doing now, they will not succeed."[59]

The fact that most of the wanaharakati were members of Chadema, an opposition party, despite the group's professed nonpartisan ideals, also did not register favorably among the majority CCM-followers in Bozi. More importantly, villagers in Matipwili were largely unaware of the wanaharakati's existence, and in turn the wanaharakati did not realize until much later in their mobilization that the EcoEnergy project also directly impacted Matipwili. Those few villagers in Matipwili who had heard about the group questioned its members' motivations. As one elder put it,

> The wanaharakati are just trying to smell things out, to see what ideas people might buy into. The way I see it, what's empowering them is the fact that they were "counted" by the project [during the PPC and hence made eligible for compensation], not that they were born and raised here like we have been. They say they want their rights

back; that probably just means they want to make sure they get their compensation.[60]

At the crux of their uneasiness with the wanaharakati was the question of whether and to what extent they really belonged to the land.[61]

I asked Samwel if he ever considered building alliances with longtime residents or recruiting them as additional members, but he seemed unconcerned and unreceptive to the idea: "When the Father of the Nation [Nyerere] was fighting for independence, he didn't go to every citizen to get his ideas across. He had a small group of people and started from there."[62] Abdallah agreed, "Not everyone is prepared to die, but we are."[63] On the subject of recruiting female members, the wanaharakati demurred. Some of them said it was "better to remain with fewer people," while others were unequivocal in their assessment that women were not "suitable comrades."[64] Samwel and Abdallah argued that women could not "die for their cause" the same way as they could, because women could not abandon their domestic responsibilities. They also raised doubts about whether women could keep sensitive information confidential and whether they might "take the group backwards."[65]

At one point, however, the wanaharakati did recruit a female member, Jennifa. She was a single mother in her late thirties who had migrated to No. 4 in 2010 from Tandika, one of Dar es Salaam's slum neighborhoods. She was outspoken in her criticism of the government and EcoEnergy. Just like the other members of the wanaharakati, she liked to appropriate the rhetoric of the ruling party to get her point across. For instance, when I asked why she was interested in working with the wanaharakati, she said, "Magufuli [in his campaign slogan] has been telling us *hapa kazi tu* [there is only work here, or we are here to work]. For us, farming *is* our work. And you cannot farm without land. Land is the economic foundation of all human beings. . . . Until the government recognizes us as human beings here, we cannot succeed in our work."[66]

Despite her desire to remain active in the group, she faced a number of challenges, not only because of her gender, but also because of her class and social status as a poor single mother. It was difficult for her to attend meetings and engage in organizing activities because she lacked financial resources, transportation, and child care support. Other constraints, such as not having enough airtime or network signal on her phone, meant that she was not able to maintain regular communication with the group, and she eventually withdrew. Reflecting on her brief stint in the group, she said, "They didn't seem to trust me. They would ignore me and go on to do their things without telling me."[67] Other women, if they were aware of the wanaharakati at all, wondered whether the group could really represent their interests or be in solidarity with them. Their misgivings about women's exclusion from politics were not confined to the wanaharakati's work; they saw it as a broader cultural problem. As one woman in Bozi put it,

> Women must serve as leaders. But from what I see, women are not trusted. Women are oppressed. Women are not taken seriously in our society because we've had less than seven years of education. Women depend on men because it's not easy for us to get together. We are overworked. Some women don't have the courage to speak up. Some feel shame about leaving their duties in the home, although men are also capable of doing them.[68]

Six months into their organizing, the wanaharakati struggled to further their agenda or inspire public confidence in their work. After Jennifa dropped out, the group did not try to bring her back or to recruit other members, female or otherwise. Not only had their membership continued to remain limited and exclusive, but internal fault lines also began to emerge. The more time I spent with the group, the more I noticed the decreasing attendance of key members like Majid, the deputy spokesperson. At a meeting in late September 2016, the group broached the subject of Majid's alleged betrayal. Emmanuel was furious as he described what he had witnessed a few weeks prior:

> Magufuli was passing through Bagamoyo recently for some event. Some people took advantage of that opportunity to tell the president that they wanted the [EcoEnergy] project. Among them were Majid and Yusuf [Zainab's husband; see chapter 4]. There were others. When the motorcade passed through the Bagamoyo-Msata road, those men held up placards that said "We want the EcoEnergy project!" "We love the project!" After that event, Majid is not even mixing words with us. There is no doubt that he and Yusuf were bought off by EcoEnergy. They have become EcoEnergy's *chachu*. I heard the company paid Majid TZS 50,000 to recruit people to hold up those placards. I heard they even promised to give him their Prado [Toyota Land Cruiser] if they succeeded in getting the project off the ground. This is making me very upset. Majid was one of our own, but now he's on EcoEnergy's side. He was a guy we trusted.[69]

Bozi residents I interviewed confirmed that Majid, Yusuf, and others tried to get the poorest villagers to hold up the placards by paying them TZS 5,000 each. Reflecting on what she witnessed, one woman bemoaned, "All the years of uncertainty have broken people here—their economy and their intellect.... People like

Majid and Yusuf know better than to act against their own people. They know that 100 percent of us do not want the project. They know that the company has not accomplished anything to date for the past ten years. They are just being played by EcoEnergy."[70]

Majid's betrayal was a fatal blow to the wanaharakati, not only to their group cohesion and morale, but also to their reputation among other villagers, which had been tenuous from the get-go. After the fallout with Majid, the frequency with which the group met decreased significantly, from almost every week in early 2016 to no meetings at all toward the end of the year. The group ultimately disbanded, although Emmanuel and Rama hoped to collaborate and run for the subvillage leadership if and when the local elections were made possible again.

What can we learn from the diverse examples of resistance I described in this chapter? Two key lessons stand out. The first is that whether people's resistance practices were as commonplace as ordinary talk, as risky as illicit activities, or as impressive as organized political activism, they were all embedded within plural, and often contradictory, moral economies. Their actions were informed not only by customary and counterhegemonic claims to land (the assertion that land belonged to those who worked it), but also their particular social identities and positions, spatial locations within the concession area, and their subjective encounters and confrontations with the project. In articulating their complaints and defending their actions and circumstances, people were making multiple moral claims. They questioned the morality of waiting, being kept in the dark about their future; the morality of displacement and violence; the morality of the mgambo, EcoEnergy, and the state; the morality of gender relations; and the morality of land use, subsistence, and agrarian citizenship.

The second takeaway is that while social difference shaped the heterogeneous channels of grassroots action, it limited the possibilities for meaningful coalition building and collective action against land grabbing. Though the wanaharakati's rallying point of agrarian citizenship was broad and flexible enough to appeal to diverse migrant populations, it was an ambiguous and nervous one for long-standing residents and women whom the group excluded. There was no unified community of agrarian citizens within the 20,400-hectare land concession, nor did the wanaharakati succeed in imagining and enacting one through their fragmented activism. Black feminist scholars have suggested that conceptualizing identities as inherently coalitional—that is, thinking coalitionally about identities—has the potential to allow people to bridge social difference and work toward building solidarity between seemingly disparate communities.[71] Had the wanaharakati forged alliances among individuals and groups with different social and spatial identities, would their outcomes have been any different? What if some elders and longtime residents took political action that they justified as being more legitimate and inclusive than the wanaharakati's? The next chapter explores precisely that eventuality, but an action that was also built on gendered exclusions, contradictions, and erasures.

6

OF PRIVILEGE, LAWFARE, AND PERVERSE RESISTANCE

> In the rich and sometimes contradictory details of resistance, the complex workings of social power can be traced.
>
> Lila Abu-Lughod, "The Romance of Resistance"

All there was left in the midst of rubble and debris was a papaya tree and a ripped CCM party flag, flying high in spite of it all. Before the demolition, there stood a CCM office in Makaani, built by the sweat and labor of all the villagers, Bambadi said, showing me a picture of what used to be. "Everyone contributed," he said. "Women and children fetched water from the river and brought it so men could use it for construction. We finished it in 2011. We used the office for holding monthly party meetings, mobilizing member support, and registering new members."[1]

The demolition came four days before the government granted EcoEnergy land title in May 2013. Bambadi and others who witnessed the event said they stood there stunned, watching the bulldozer crush the building to its foundations. The district commissioner and a team of district party officials supervised the demolition, under the directive of the minister of lands, Anna Tibaijuka, an early supporter of the EcoEnergy land deal (see chapter 1).[2] Bambadi was adamant that EcoEnergy, too, was complicit in abetting the demolition: "It was all secretive. EcoEnergy gave about two hundred million shillings to the district commissioner to abolish the building. There is no paper trail on this, and of course I don't have evidence of that, but ask anybody. We know it is true."[3]

A year prior, Bambadi and two other local elders, representing themselves and over five hundred supposed local residents in Makaani and Gama, had filed a trespass lawsuit against EcoEnergy, the Bagamoyo District Council, the commissioner for lands, and the attorney general at the High Court of Tanzania. According to Bambadi, the demolition was another indisputable evidence of trespass, in addition to the land survey and the PPC the government conducted in 2008 and 2011 respectively on the plaintiffs' land without their consent. In his view, the demolition represented a gross violation not only of their land rights but also of their civil and political rights. For prominent ruling party politicians to voluntarily destroy the home of a local party cell (*shina la wakereketwa*)—the most basic electoral and political mobilization unit in the countryside—was an outright rejection of rural citizens, a statement that underscored that neither their voices nor votes mattered. The only way to reclaim their rights and recognition, Bambadi said, was to fight back in ways that were legible to the state, or by the using the weapons of the powerful against them: "The government is all about the law. If they care about the oppressed, they should be protecting our rights, not the foreign investor's!"[4]

Between 2013 and 2014, the national news media reported on the pending court case and the demolition with headlines such as "Villagers Cry over Land Grabbing" and "Villagers Up in Arms for Fear of Losing Land."[5] In its global campaign report released in early 2015, ActionAid International described the lawsuit as resulting in part from EcoEnergy's failure to obtain free, prior, informed consent from the local communities.[6]

Against these public condemnations, however, the High Court of Tanzania dismissed the case in November 2015, citing the plaintiffs' failure to provide evidence to support their claims. When I interviewed Bambadi soon after the verdict, he was emphatic that their lawsuit was rightful resistance to land grabbing and that they were determined to carry on with the litigation: "We will appeal. We deserve to be on the land. We are rightful farmers, residents, and citizens. It's not right that the investor wants to own all of our land."[7]

The elders' lawsuit appears at first glance as an example of formal-legal resistance—a repertoire of contentious politics where subordinated individuals and groups harness the law, judicial processes, and rights-based discourses to contest perceived violations of justice and demand fair redress.[8] Recourse to law, of course, is not new to land politics in Africa. Legal contention was at the heart of indigenous struggles against colonial and postcolonial land annexations, as documented by the well-known Meru and Barabaig Land Cases in Tanzania.[9] Early leaders of the African National Congress in South Africa, too, had turned to the court system to challenge and claim recompense for the mass dispossession of Black people from their land.[10] And recent years have seen the emergence of strategic litigation as a response to land grabbing, initiated by local actors in close alliance with transnational NGOs, social movements, and legal experts.[11] This "judicialization" of land deal politics may signal expanding and more complex avenues of grassroots struggle, as Ruth Hall and colleagues have contended.[12] Others have suggested that while efforts through the courts to resist land grabbing may not always be effective, they can be instrumental in shaping public discourse around land not as an object of business transactions, but as an issue of human rights and identities.[13]

My own evidence suggests, however, that what resembles formal-legal resistance can sometimes obscure the complex gendered power relations and local political dynamics that underlie struggles over land. While not denying or devaluing different expressions of subaltern agency, this chapter asks who the male elders are, who was included and excluded in the lawsuit, why, and with what consequences? Here I draw on the insights of feminist anthropologists who have long insisted that critical scholars refuse the impulse to "sanitize" the multifaceted interests, intentions, and internal politics of subordinated groups and instead use resistance as an opportunity to engage in an ethnographically thick "diagnostic of power."[14] As I will show, the lead plaintiffs in the present case were able to draw on their multiple social positions of privilege—as men, elders, husbands, long-term residents, village leaders, and rural party elites—to pursue legal action, but in ways that appropriated, erased, and transgressed the land rights of diverse other legitimate resource users, including, most immediately, their wives.

To avoid conflating the male elders' lawsuit with an innocuous form of resistance, I shall characterize it as an expression of "lawfare" from below, or the use or misuse of law by elites among the poor to seize opportunities for personal gain at the expense and exclusion of many.[15] It is seldom the weakest or the most marginal that predominate in lawfare from below. This is perhaps unsurprising, given how uneven access to justice has historically been in Africa and elsewhere around the world. Not only does litigation require considerable investments of time, money, and political resources, but also it has been rare for African courts to yield favorable results for rural citizens.[16] On the rare occasions when rural people are able to lodge legal claims, the plaintiffs are likely to be privileged actors, especially male elites in places where patriarchal authority over land prevails. Considering the prohibitive costs, dismal historical record, and the general lack of confidence and trust in the court system, rural people have little incentive to bring legal charges against governments and investors even when they feel wronged or when their livelihoods are at stake. It is likely for these reasons that recent cases of litigation against transnational land grabbing, as noted above, have invariably involved the support of extralocal allies like transnational advocacy groups that have the necessary resources, visibility, and reach. In contrast, the allies of the plaintiffs in the present case have remained unnamed. By attending to not only what was said in the courtroom but also what remained unsaid, this chapter illuminates the operations of gender power, privilege, and inequality that shaped the making of the lawsuit and the ramifications of its premature dismissal.

Land Case No. 126 of 2012

We begin with the facts of the case. In their complaint submitted to the High Court of Tanzania (Land Division) in Dar es Salaam on August 16, 2012, the plaintiffs—three male elders representing themselves and 537 others—made the following claims and allegations: (1) that they were lawful owners of various plots of land in "Makaani Gama village" measuring a total of six thousand hectares; (2) that the government's surveying and designation of the land in dispute for investment purposes was null and void; (3) that the government's granting of the certificate of title to EcoEnergy was illegal; and (4) that the plaintiffs deserved to receive TZS 4 billion (roughly USD 2 million) in compensation for economic losses due to the defendants' trespass, in addition to payments for general damages and other relief as determined by the court, plus any interest on those payments; and (5) that they should also be compensated for the costs of the lawsuit.[17]

Court documents used the names "Makaani," "Gama," "Gama-Makaani," and "Makaani-Gama" interchangeably and described them as a village. It needs clarifying, however, that neither Gama nor Makaani, nor the two areas combined, is a registered village or an administrative unit. At the time of the lawsuit, Gama fell within the administrative boundaries of Kitame subvillage. When it came to Makaani, however, there was confusion among people whether it belonged to Kitame or Razaba subvillage (both within the larger Makurunge village/township). Though Makaani residents I interviewed said they would go to Razaba for village meetings, political rallies, or to receive rare social service benefits like mosquito nets from the district government, the High Court considered Gama and Makaani to be part of Kitame for the purposes of adjudicating whether the plaintiffs were customary holders of village land. As chapter 2 discussed and as I return to later, these administrative boundaries are relatively recent constructs; for people who have lived through the multiple boundary redrawings since colonial times and who have been farming in the dynamic floodplain landscape, these borders seemed irrelevant to their everyday practices.

Among the charges outlined above, the first point needs further elaboration not only because it was at the heart of the lawsuit, but also because the plaintiffs' land claims were highly heterogeneous. The first plaintiff, Bambadi, less often known by his full name, Salum Yusuf Salum, claimed that he was born in Gama in 1946 and raised and educated there, and that he had inherited three hundred acres in the area from his deceased parents. The second plaintiff, Ally "Thabiti" Ngwega, adduced that he owned a ten-acre farm plot in the disputed area, which he had inherited from his father in 2007. The third plaintiff, Ally Said, similarly stated that he had inherited eighty-nine acres from his deceased parents and acquired an additional thirty-one acres in the disputed area by clearing pristine land. The rest of the plaintiffs argued that they owned various plots of land in the disputed area, ranging from ten to thirty acres, which they claimed to have acquired by clearing bushland between 1986 and 2007 before EcoEnergy's arrival.

In a verdict rendered on November 13, 2015, a High Court judge dismissed the plaintiffs' claims with costs, meaning they not only lost the lawsuit but were liable for compensating the defendants for wasting their time and money. The fundamental reason for the dismissal was that the plaintiffs had not provided sufficient and credible evidence to support their claims. The Village Land Act of 1999 adjudicates customary land rights based first and foremost on government-issued land titles or certificates of customary rights of occupancy. In the absence of these documents, the plaintiffs could submit other forms of evidence, including oral testimonies or affidavits stating that they have been in "peaceable, open, and uninterrupted occupation of village land under customary law for not less than twelve years," or that they have been using the land regardless of the number of years under an arrangement or transaction that is evidenced in writing.[18] In the present case, only 46 out of the 540 plaintiffs had submitted evidence to prove their customary landownership. Moreover, no one had submitted any clear breakdown of the said damages, worth TZS 4 billion. Given that the six thousand hectares in question were not jointly owned by the plaintiffs and that each claim was independent and heterogeneous, the judge, drawing on precedents, deemed that it was the duty of each appellant to testify and prove their respective landownership as well as the specifics of the losses incurred. The plaintiffs' lawyer offered no rebuttal in response to these points.

That the plaintiffs failed to provide adequate and sufficient evidence and, more importantly, that their lawyer allowed them to proceed with the case despite this critical omission is curious, if not problematic. Beyond this, there were other issues that could qualify as attorney misconduct. The judgment indicates, for example, that one of the witnesses who testified in court was a "legal officer" from the law firm that was representing the plaintiffs. My interviews with the wives of the lead plaintiffs, Bambadi and Thabiti, revealed that their lawyer also had a stake in the lawsuit, resulting from his direct land purchase from none other than these elders—a circumstance undisclosed in the judgment. It is unsurprising, then, that the case was dismissed with costs. A prominent Tanzanian legal scholar whom I consulted suspected that the case had been "set up to fail."[19] He adduced this from the irregularities described above and from the fact that the case took only about three years to come to a verdict, whereas typical High Court land cases can take ten years or more to reach resolution.

EcoEnergy was almost certain that the lawsuit was a ploy orchestrated by special interest groups, namely the country's sugar importers, who risked becoming redundant in the face of the government's increased support for import substitution industrialization of sugar.[20] As the company chairman, Per Carstedt told me, "The [sugar] traders are doing everything they can to obstruct, frustrate, delay, and stop the project. By delaying the project more and more [with the lawsuit], you end up creating more problems for the project."[21] Carstedt speculated that these industry players co-opted the male elders, because they were already well-known opponents of the land deal. Other commentators have also hinted at the role of sugar importers in thwarting the EcoEnergy project and other sugar schemes in the nation, although evidence has been difficult to establish.[22] When I asked the elders how much the lawsuit had cost them and how they were able to afford it, they were equivocal in their response, only to say that the fees amounted to several hundred million shillings and that they received support from their fellow villagers and unnamed well-wishing "sponsors."[23]

Besides the partiality of evidence, conflict of interest, and the ambiguity of motives of different actors behind the lawsuit, another important issue remained unaddressed during the litigation: the absence or exclusion of the plaintiffs' wives and hundreds of other legitimate land users from the lawsuit. To investigate the reasons for and implications of this omission, I provide below a more in-depth look at the plaintiffs' biographies and their historical relationship to the land. I focus on Bambadi and Thabiti here because of their leading roles in the lawsuit and because, during my fieldwork, the third plaintiff was absent from Bagamoyo owing to illness; according to court documents, he did not testify in court or offer any evidence to support his land claims.

Bambadi

Bambadi was a tall, loud, and formidable man in his early seventies. With "-*bamba*" meaning "to arrest, catch, or hold," his nickname implied he was a person of notoriety. According to elders in Matipwili village, Bambadi gained the nickname in the 1980s when he was convicted of embezzling funds from the village treasury during his tenure as a subvillage chairman of Kisauke. As we saw in chapter 2, Kisauke had been part of Matipwili until it was annexed by Saadani National Park in the early 2000s. The evicted Kisauke villagers, including Bambadi, Thabiti, and their families, initially moved to Gama in the floodplains, where they traditionally farmed; several years later, they resettled and built new homes on higher elevation in Makaani, a few kilometers south of Gama. The relevance of this detail will become evident later when I introduce Thabiti in the next section.

Bambadi was not fazed by his roguish epithet. During our first interview in August 2014, he introduced himself, without hesitation, as Bambadi, laughing boisterously and beating his chest with pride. Many women and the youth in the area regarded Bambadi with a mix of deference and fear, by virtue of his gender, seniority, and political status, as well as his fiery temperament. He was almost always seen in company with other men, often wearing a green-and-yellow cap and T-shirt signaling his leadership role in the local CCM party cell (figure 6.1). While some people thought he genuinely cared about community development in Gama and Makaani, many more contended that his primary goal in politics was to accumulate power and wealth. Matipwili villagers who had known Bambadi for many decades also doubted his professed loyalty to the CCM. As one villager noted, "Bambadi is just using *ujanja ujanja* [tricks, deceit]. You know, Bambadi and his friends have membership cards for more than three political parties. They plant CCM flags in front of their houses, but that is just a cover."[24]

Even if they had never met Bambadi in person, everyone I interviewed, from Bozi to Matipwili, knew of him (and by association, Thabiti) because of his shady land sales to various "strangers" at the (now demolished) CCM office in Makaani. People speculated that he was selling land to outsiders to increase the total number of bodies occupying Makaani, so as to make it difficult for EcoEnergy and the government to clear the land. "Imagine journalists taking pictures of government

FIGURE 6.1. Bambadi speaking to the crowd at a meeting in Makaani. Photo by Selemani, December 2015.

bulldozers running over poor rural farmers. That will surely set things on fire!" said a young man in Makaani who was both critical of and empathetic to Bambadi's suspected motives.[25] The timing of when he began selling land supposedly coincided with the PPC in 2011, which he had rejected (see chapter 3). As one elder in Matipwili told me, "Bambadi refused to be counted during the government valuation [PPC]. I think that's when Bambadi started selling land. I heard he got someone to post advertisements on the internet. That's how he was able to recruit five hundred people for the court case. I live here, but I don't know those people. Bambadi knows where they live."[26]

What made Bambadi's transactions particularly contentious was that he was selling land not only in Gama and Makaani but also across the Wami floodplains, even up to the coastline. "We fought so hard with Bambadi when he was trying to sell our land to rich people in Dar es Salaam. And then, EcoEnergy came and tried to take our land!" a female floodplain farmer lamented.[27] On the whole, people saw Bambadi as appropriating his de facto leadership position in Makaani, as well as his presumed moral authority linked to his seniority, long-term residence, and local knowledge of the area, to justify his land sales.

Bambadi's clientele for land sales comprised two main groups: poor landless migrant farmers and wealthy urban elites. The first group was presumably an easy target, considering the continued influx of newcomers into the area since around 2010 when news of the EcoEnergy project spread. Most migrants in Makaani admitted to either having bought or been given land for free from Bambadi. For example, a fifty-three-year-old male farmer who had migrated from Iringa noted that when he arrived in Makaani in 2011, the first thing he did, encouraged by word of mouth, was to visit Bambadi to introduce himself and request an allocation of land. When I asked why he did so, he said, "I didn't know who he was, but he was an elder, and people said he was the local leader. As long as he is an elder, a newcomer like me just assumes he knows about land matters around here from the past. He was pleased I was coming to the area to farm and gave me 2.5 acres for free."[28] While Bambadi gave this man and several other young male migrants land for free as long as they cleared and used it, others, including single mothers, divorcées, and widows like Rukia from chapter 3, said they had to pay Bambadi a fee of anywhere between TZS 2,500 and 5,000.

Bambadi's land sales to wealthy urbanites were more ambiguous and difficult to trace, but they were made evident to me one afternoon when I was visiting with him. He had invited me to come discuss the High Court verdict. As we got seated around a plastic table outside his house, he told me he was going to appeal. But before I could ask further questions about his plan, we were interrupted by a small white truck approaching from the direction of Bagamoyo town. The truck pulled up in front of Bambadi's house. A middle-aged woman in modern attire with sleeveless shirt and smart trousers, shiny jewelry, a shoulder bag, and platform sandals, accompanied by a young, equally well-dressed male driver, got out and walked toward us. Their visit seemed unexpected. Bambadi got up from his chair and unceremoniously introduced the visitors to me as *wenyeji* (local people / natives of the area), and me to them as a *mtafiti* (researcher). The woman smiled and said she owned a watermelon farm in Gama and that she was just passing by to say hello to the elder. In an effort to make small talk, I asked whether she was one of the plaintiffs in the case. As she smiled and nodded, Bambadi quickly intervened: "When you cheat people, you get followers. And when you have many people who feel cheated, it is easy to get followers." After what felt like an awkward exchange, Bambadi escorted his guests inside the house, leaving me alone by the table.[29]

Hadija, Bambadi's wife, motioned me to join her in the kitchen, where she was making a big pot of maize *ugali*. Bambadi and the visitors reemerged shortly afterward, got into the truck, and left, saying they needed to do "farm visits." Hadija continued boiling the water for *ugali*, and when the truck was out of sight, she told me, unsolicited, that the visitors were from Mbezi, a neighborhood in Dar es Salaam, and that she had seen them before when they came to "look for land," revealing that Bambadi had brokered the land sale. I took the rare opportunity to interview Hadija in the absence of Bambadi. However, after a series of basic questions, it became apparent that Bambadi had shared little information about the court case with his wife. Perhaps exasperated by needing to repeat "I don't know," Hadija explained,

> If I look around, women around here are losing a lot. Women are invisible. I don't really know why . . . maybe because we haven't studied. But even if we have studied, we are being cheated. Women are being oppressed by their husbands, brothers-in-laws, and etc. I am a Zigua [a historically matrilineal-matrilocal ethnic group] from Mkange. . . . Here comes the problem. I came all the way here from Mkange for marriage. Bambadi is a Doe [a historically patrilineal-patrilocal ethnic group], so I had to come live with him. But here, I have no relatives. I have to rely solely on my husband, including land. Even if I get a divorce, I cannot force him to divide the land equally. . . . It's like women are owned by men, and men are responsible for everything.[30]

As her quote suggests, while Hadija was hardly involved in the court case, she was well aware that her access to land critically depended on her marriage to Bambadi and the lawsuit. When they got married, Bambadi had given her permission to grow food crops and vegetables on eight acres behind their house, although she could only farm at most two acres by herself. She was unsure how much land

Bambadi owned or claimed that he owned, but she indicated that he operated a "big farm" in Gama with hired laborers, which was not an option for most poor smallholders in the area.

In his oral testimony in court, Bambadi was vague about when and how he came to inherit the three hundred acres from his parents. He simply stated that he was born, raised, and educated in the area and that his parents were laid to rest there. According to local elders, it was impossible for Bambadi, or for any villager for that matter, to have inherited a tract of land as vast as three hundred acres. Indeed, the median landholding per household among the 176 households I interviewed within and immediately outside the EcoEnergy concession area was three acres (a little more than a hectare). Given the absurdity of his alleged land claims, other elders in the area, especially those in Matipwili familiar with Bambadi's track record of misdeeds, distanced themselves from his court case. On his part, Bambadi also did not actively seek their support or involvement, arguably because he knew it would not help the case to have so many people who were capable of questioning and contradicting his claims. The next section adds more historical complexity to the origins of the case, while shedding further light on why exclusionary politics remained so central to the male elders' lawfare.

Thabiti

Thabiti, the second plaintiff, was more patient and calmer in his demeanor than Bambadi. He was two years older than Bambadi and was almost always seen wearing an Islamic *kanzu* (long-sleeved full-length gown) and *kofia* (embroidered cap), rather than the green-and-yellow CCM attire.

Like Bambadi and Hadija, Thabiti and his wife moved from Kisauke to Gama when Saadani National Park annexed their subvillage in 2003. At the time of their eviction, Thabiti was the chairman of Kisauke. The move to Gama was meant to be temporary, Thabiti said. Although Gama was an ideal place for farming, its low-lying topography and the hydrology of the river meant that it wasn't a suitable place to build permanent homes. As noted in chapter 2, farmers on the south side of the river have historically dwelled in areas of higher elevation to avoid the risk of water damage from the annual inundation of the floodplains.

According to Thabiti, the evicted villagers, many of whom had their farms in Gama, had two options. The first was to move to higher ground near the Matipwili village center on the north side of the river. This wasn't a popular choice, however, as it meant they would have to pass the national park gate and confront the guards there to reach their farms in Gama. The second option was to resettle in Kitame subvillage on the south side of the river, close to the Indian

Ocean. This option would have obviated direct confrontation with the national park guards, but it also wasn't a popular choice for many, because the road from Gama to Kitame was unpaved, and the heavy clay soil that turned to sticky mud when wet would make traveling back and forth to their farms extremely difficult.

On November 30, 2003, Thabiti convened a meeting with the affected villagers to propose a third option, of moving to Makaani. Though it isn't clear from the minutes retrieved from the Matipwili village archives how many villagers attended the meeting and how they responded to this proposal, Thabiti, together with Bambadi, went on to make a number of demands to district government officials, local politicians, and the national park authorities.[31] First, they insisted they had the right to choose wherever they wished to resettle, as well as the right to compensation for their losses. Second, they requested to resettle and establish a new subvillage in Makaani; this subvillage would fall under Makurunge village (similar to Kitame and Razaba subvillages). Third, they demanded that the government provide Makaani residents with social services, housing, health care, education, and water. Lastly, they asked the government to survey and officially registered the new subvillage. Thabiti also forwarded these demands to the Makurunge chairman.[32]

Five years on, however, Thabiti had not heard back from any government leaders. Some of the evicted Kisauke residents like himself remained in Gama, despite the trouble of having to repair their homes regularly and take refuge during times of heavy rainfall and river flows. Others ended up moving to Kitame or up north to Matipwili, notwithstanding the challenges mentioned above. In 2008, the sixth year after their eviction, Thabiti said he received a surprise visit from a team of land surveyors sent by the minister of lands. "I thought they were coming to give us permission to settle in Makaani," Thabiti said. The purpose of the land survey, however, was to apportion the former RAZABA ranch for Sekab's biofuel investment (see chapter 1). "When we found out it was for Sekab [later EcoEnergy], I was asking myself and my fellows, 'Have we been forgotten?'"[33]

In December 2009, after an unusually heavy rainfall, Thabiti wrote once more to the Makurunge chairman to request permission to resettle in Makaani, citing the severity of floods.[34] Again, he did not receive a response, but this time he was undeterred. On January 17, 2010, he mobilized his fellow villagers to occupy and resettle in Makaani. Soon thereafter, Thabiti and Bambadi hatched their plan to establish a new CCM party cell there, but without approval from higher levels of party organization, according to the CCM Kitame branch chairman. Thabiti was allegedly elected chairman of the party cell, and Bambadi the chairman of their supposed new "subvillage." On March 10, 2010, Thabiti and Bambadi sent a letter to the Bagamoyo district commissioner, signed by 150 so-called "residents of Gama (Makaani)"—or what they interchangeably referred to as "Makaani

(Gama)"—to insist on their rights to establish permanent residence in Makaani. Whereas Thabiti's and Bambadi's original request to government leaders in 2003 had emphasized their right to resettle in Makaani as a result of their forced eviction by the national park, the letter and the petition they sent in the post-2009 period hardly mentioned this critical historical fact and instead focused solely on the recent flood damage.[35]

Throughout 2010 and 2011, Makaani residents received multiple notices from ward-level authorities, the RAZABA manager, and from CCM leaders of the Kitame branch to vacate the land and to close down their underground party cell. When they failed to heed these orders, the RAZABA manager demanded immediate intervention from the district commissioner to resolve the "excessive land invasion" and illegal land sales happening both in the former state ranch and part of the ranch that was subdivided to Sekab:

> It is said that plots there were sold at TZS 3000/acre and a large part of the area has already been sold to visitors coming from outside Bagamoyo, most of the buyers coming from Dar es Salaam. The effects of that land invasion can be seen through deforestation as well as settlements along the road inside the ranch. I am submitting this report to you for proper actions to be taken before this conflict reaches a point of controversy.[36]

The following month, in December 2010, the district commissioner visited Makaani, and there he ordered all residents to leave the area within one week and "go back to where they came from." When people refused to do so, he arrested and charged their stated leaders, Thabiti and Bambadi, with criminal trespass on government property.[37] What transpired with the charges remains unclear, but interviews with anonymous informants suggest that Thabiti and Bambadi were bailed out soon after their arrest by some unknown supporters in Dar es Salaam.[38] Once released, the two elders began planning their lawsuit and the construction of a CCM office, which they called the "Gama branch," arguably in contempt of the leaders at the Kitame branch. The building reportedly cost them around TZS 40 million (USD 20,000), which was a hefty sum to raise among the predominantly poor smallholders in the area who did not have regular sources of income. When asked how they raised the funds, Thabiti and Bambadi, like their previous response to my question of how they funded the lawsuit, vaguely referred to the contributions of local villagers and the support of "well-wishers."[39] Like Bambadi, Thabiti explained that they used the CCM office primarily for registering new members to the party and mobilizing their support toward various community development goals like building schools and wells. Yet many local residents, including those who had followed Thabiti from

Kisauke to Gama to now Makaani, suggested the office was used exclusively by the two elders for selling land.

While it was more common for people to point the finger at Bambadi for engaging in illegal land sales, Thabiti was very much complicit in the act, as his wife Fatuma affirmed. Like Hadija, Fatuma could say little about the origins or the details of the court case, but she was cognizant of its covert nature, as she made clear:

> I used to ask [Thabiti] about the case, but he would hide his papers. So did Bambadi. But from what I know, the five hundred people who joined the case were from Dar es Salaam. They were given land by Thabiti and Bambadi. Oh! Even the lawyer bought a farm here, although I have never seen him use it! [*Laughs*] The lawyer came here maybe twice. It is mostly Bambadi and Thabiti who are called into meetings with the lawyer in Dar [es Salaam]. Sometimes at night, when my husband's phone rings, I lie there wondering whether someone has died, but it is nothing like that. He is just being called into another meeting in Dar with the lawyer or with someone else. It would have been good if the lawyer and judge had come here to talk to me. I don't know if any women here are involved.[40]

Fatuma's testimony is important for at least two reasons. First, she highlights that the 537 co-plaintiffs who joined Bambadi and Thabiti were not actual local residents. In fact, judging by the names of plaintiffs who submitted affidavits of landownership, a good number of them were of Chagga origin, one of the wealthiest and the most educated ethnic groups in Tanzania, and unlikely constituents of rural Bagamoyo. Second, when asked to elaborate on why she wished the lawyer and the judge had spoken with her, Fatuma disclosed information that was contrary to what Thabiti had said in court. In his oral testimony, he stated that he had inherited ten acres of farmland from his father in 2007. Unlike Bambadi, who had appealed to his birthright to shore up his customary land claims, Thabiti abstained from sharing with the court any personal details about his life. Neither did he wish to discuss the subject with me during interviews, except to say that he was born in 1944 and that he has lived in the area for a "long time."

According to Fatuma, however, the ten-acre farm in Gama that Thabiti claimed was his was in fact hers. It was land that she had inherited from her parents, who passed away in the 1950s. Her father was a migrant laborer who had worked in various coastal colonial plantations from Lindi to Tanga, until he finally settled in Bagamoyo in the late 1920s. I quote her story at length below because she demonstrates how places and their meanings come to matter through complex entanglements of history, culture, and power. In so doing, she highlights the limits of customary land claims that are based solely on the labor theory of property adduced by most plaintiffs.

> My father was a Makua born in Masasi when the Germans came to rule Tanganyika. After finishing school, he moved to Lindi and worked at a salt mine as a migrant laborer. Later, his boss transferred him to Kunduchi in Dar [es Salaam] to work at a sisal plantation. My maternal uncle worked there at the same plantation as a messenger, delivering letters.
>
> One day, my mother went to visit her brother in Kunduchi. When my father saw her, he was immediately smitten. He asked people, "Who is that girl? To whom does she belong?" These things of men.... [*Laughs*] When he found out that she was related to the plantation messenger, he asked my uncle, "I want to marry your sister." My uncle replied, "I don't have authority to give her to you. Go to my father in Tunduru." So he did. My mother was very young when she got married, maybe fifteen or younger....
>
> After a while the *wazungu* [Europeans] transferred my father again from Kunduchi to another plantation in Muheza in Tanga. My father had a three-year contract there. After three years, he was transferred to another plantation within Muheza, called Kicheba, which white South Africans owned. Our family spent many years in Kicheba. My mother gave birth to eight out of twelve children there, including me. But my father lost his job when the South Africans went back to their country.
>
> My father had three maternal uncles at that time living in Bagamoyo. These uncles were originally hunters who were sent from Masasi to Bagamoyo to hunt elephants for the Arabs. This was a long time ago. When they first came to this area, they decided to call it Makaani, which in their vernacular meant "to stay," like the Swahili *kaaeni*, which means "let us stay." After a while, they decided to move closer to the Wami River, where it was better for growing food.
>
> When my father lost his job in Kicheba, his uncles urged him to come visit. They were unhappy because my father had stayed in Tanga for far too long; they thought it was better for family to live closer. My father was worried: "But how can I [move]? I have my family and coconut trees here in Tanga." But he could not disobey his uncles' wishes. When he eventually moved to this area around the late 1920s, he got a job at the sisal estate in Kisauke. This is before Saadani [National Park] took Kisauke. My father worked there for a long time.... After a few years of working there, he decided to bring his wife and children from Tanga. The uncles gave my parents ten acres to farm in Gama. Those areas

[Kisauke, Gama, Makaani] were known as Wami back then. . . . Later, when my parents died, they gave the land to me. I know that it is my parents' land because they planted a tall *mvumo* [African fan palm] there, which still stands to this day.[41]

The mvumo was indeed a local landmark (see figures 6.2 and 6.3). People readily identified it with Gama and the "Thabiti's land," a shorthand that inadvertently erased Fatuma's land rights. In 1972, Fatuma married Thabiti, who was then a contract laborer at a British-owned salt mine in Kitame to the northeast of Gama. She recalled that at the time of their marriage, Thabiti was a migrant worker with no land or family ties in Bagamoyo: "He is a Pogoro from Morogoro who came to Bagamoyo for work. His parents visited once in the late 1970s, but neither of them lived long after that." Because Thabiti had no access to farmland of his own, he relied on Fatuma's inherited land and her labor to provide food for the family once they married.

For Fatuma, the entwined history of her family and colonialism was inscribed in the landscape. Moving to Makaani meant "returning," as she described, "to

FIGURE 6.2. The road from Makaani to Gama during *masika* (the long rainy season). The *mvumo* (African fan palm) stands at the location of Fatuma's and Thabiti's farm. Photo by Hasani, February 2016.

FIGURE 6.3. Gama during the long rainy season, showing inundated floodplains. Photo by Hasani, February 2016.

the land of our ancestors."[42] Despite the centrality of Fatuma's story to Thabiti's land access and for understanding the historical relationship between Kisauke, Gama, and Makaani, she was excluded from the lawsuit, and so her testimony never made it into the courtroom. Her entitlements were rendered invisible, or at best misappropriated by and subsumed under the authority of her husband. The judge and the lawyers on both sides of the case never recognized her existence and land rights—let alone those of Bambadi's wife and numerous other legitimate landholders. Yet what is not said is as critical, if not more critical, to understanding how power works. On the one hand, the elders' lawfare and the exclusionary practices that undergirded it demonstrate the everyday operations of elite male power at both the household and community level. On the other hand, the failure of the lawyers and the High Court judge to consider the possibility that the elders might have family members with separate, independent land claims reflects what feminist scholars and lawyers in Tanzania have problematized as the patriarchal disposition of the judiciary as a key arena of state power.[43]

The Aftermath

The news of the High Court ruling traveled fast. In February 2016, EcoEnergy distributed a one-page circular to various communities within the concession area, including and beyond Gama and Makaani. Some people found the memo pegged to a tree, while others received individual notices firsthand from their local leaders or secondhand from their neighbors. Some described how Ali, the young Tanzanian EcoEnergy employee, came on a motorbike and threw the papers into the air as if to celebrate the outcome. Summarizing the verdict, the memo stated that "Gama-Makaani" was the legal property of EcoEnergy. That much the local people understood, though still contested. What gave them trouble was the remainder of the notice. It declared that the plaintiffs in the lawsuit, as well as "all persons living within the project area," were "intruders" and that they were "expected to take proper action according to the law and decision of the court." That is, all current residents in the planned EcoEnergy project site, regardless of their involvement in the lawsuit, were expected to voluntarily leave the land, presumably without the earlier promised compensation, and resettle elsewhere to avoid forced eviction.

Shortly thereafter, the district commissioner halted a local election that was scheduled to take place in Razaba subvillage (see chapter 5). Neither the candidates nor the ward councilor was given clear reasons for the suspension, although they speculated it had to be related to the lawsuit. People were outraged by their sudden and unwitting disenfranchisement, and they quickly blamed Bambadi and Thabiti for diminishing their status and reducing them to mere squatters. A thirty-three-year-old male farmer, born and raised in Bozi, said angrily in an interview,

> The reason we are now called intruders is all because of the court case. But we were not involved in the case! And I have been living here for the past thirty years! Everyone here knows that those five hundred people who supported the *wazee* [elders, Bambadi and Thabiti] are from Dar es Salaam or *wageni* [strangers/migrants] to this area. It's like this, sister—you see, one fish is rotten in a bucket; that one fish is bad enough to make all the other fish look rotten! Back then when EcoEnergy counted us and promised compensation, I was skeptical. I didn't think they would pay us, but they kept promising, so I started to build some hope. Now, I don't know if that's even a possibility.[44]

Similar concerns were raised by other younger residents, who felt the elders had cut short their prospects and possibilities on the land. A thirty-nine-year-old migrant and single mother, for example, expressed her anxieties about remaining on the land and regretted having believed in Bambadi and his moral authority as a male elder:

> When I first came here in 2009, I couldn't send my eight-year-old daughter to school because there wasn't a school in Makaani. Bambadi promised young mothers like me that he would bring a teacher into our community. Later he said we each had to pay TZS 12,000 per family to pay the teacher, which I couldn't afford. I thought he was doing the court case for our community development. But now I'm not sure what his intentions were, and I don't know what will happen to my life now. I am scared. I am just waiting for things to pass and get better.[45]

To many, the elders' loss did not seem surprising. What people were aggrieved by was how they were inadvertently disenfranchised and dispossessed in situ as a result of the elders' lawfare, which they had opposed, were excluded from, or knew little about from the beginning.

Hadija and Fatuma were equally dismayed by the court's decision. Soon after the judgment was made public, I asked them how they thought the verdict might affect their lives. Hadija first gave a nonchalant shrug and said it had become "normal to live with uncertainty."[46] But soon after, she pondered whether it would be better for her to divorce Bambadi and return to her home village where her relatives lived and where she knew she had some inherited land, though it had been a while since her last visit. Yet, upon realizing she could not divorce Bambadi of her own accord, without his consent, she said half jokingly and half serious: "If there is another woman who wants to take my husband, she is more than welcome to!"[47] Fatuma, on the other hand, regretted again how she could not tell her side of the story. Rubbing the pronounced edema on her right foot, she said, "Thabiti did not even ask me if I wanted to be involved in the court case. I have a lot of stories to tell. But I have trouble with my feet. Sometimes I can't even feel my feet, as if they are paralyzed. To get from here to there [*pointing to a few feet away*], I need to stop two to three times. I wish the judge had come and spoken with me."[48]

When asked what they would have done differently if they were in positions of power and could take legal action, Fatuma and Hadija noted they would have taken on Saadani National Park instead of or in addition to EcoEnergy. Other evicted villagers from Kisauke said they would do the same but without involving Bambadi and Thabiti. While no record exists of compensation, they claimed that TANAPA had paid Bambadi and Thabiti more than TZS 10 million back in 2003 to instigate people to move. "They were paid at night. If things happen in the dark then it must be something dodgy, especially if there is already an established government practice for compensating citizens," said one former Kisauke villager who was not involved in the court case.[49] The Matipwili village chairman also found it strange that the elders had not legally pursued the national park. He did not know whether the national park had compensated them under the table but said, "It is very clear. The real enemy of the people is TANAPA. They are always tricking people. They told the villagers [in Kisauke] that they would be able to keep their land, but then went on to demarcate their own boundaries beyond the river without their consent."[50] The Kitame subvillage chairman couldn't agree more: "If Thabiti and Bambadi wanted justice, my view is that they should have gone after TANAPA. But they tried to make some money for themselves while leaving everyone behind in confusion and chaos."[51]

The fact that a few farmers in Tanzania managed to sue the national government and a foreign investor to cease the enclosure of their land is especially significant in light of the general reluctance rural people have in using the court system to access their rights. For those who decide to pursue legal action, the allure of law may be that it appears to provide "a ready means of commensuration" or a set of more or less standardized tools and practices for negotiating divergent, competing, and at times impermeable claims.[52] The way that the plaintiffs justified their case echoed this idea of commensurability; they contended they were using the court system to resist what they perceived to be unlawful trespass on their land by powerful actors.

Yet this semblance of benign legal resistance belied many contradictions, slippages, and exclusions that simultaneously enabled and undermined the lawsuit. Ambiguities still remain about the plaintiffs' motives and those of their undeclared allies. Nevertheless, it is clear how patriarchy, privilege, and inequality shaped, and were reinforced by, the male elders' lawfare. As I showed, the plaintiffs mobilized their multiple positions of power—as male elders, husbands, long-time settlers, and grassroots political leaders—to exploit resources from external actors and to exclude from the litigation diverse legitimate resource users affected by the EcoEnergy land deal. The lawsuit, understood as lawfare, was not only perverse in its design and outcome but also in its unintended consequences. The plaintiffs may have been successful in delaying the implementation of the land deal while the case remained pending, but this delay and the ultimate failure of the lawsuit rendered local people—even if they had nothing to do with the case—more vulnerable to dispossession and disenfranchisement. The restrictions on civil and political rights and the threat of eviction that followed also laid bare the marginality of rural people vis-à-vis the state and the foreign investor. They were forced to negotiate their subjectivities as rightful landholders and citizens on the one hand, and externally imposed identities as squatters and intruders on the other—all because of, to borrow from the interviewee above, a few "rotten" elites.

Would the outcome of the lawsuit have been any different if the plaintiffs were more diverse and inclusive in representation, if the case had targeted the legality of the land annexation by Saadani National Park, if witness testimonies centered more on people's lived histories on the land, and if the case was prepared more robustly by an attorney with no conflict of interest? Perhaps. But it would be naïve to ignore the violence inherent in the law; access to law is unevenly distributed, and the dispersal of land politics into the legal sphere does not guarantee justice or redress. And more law does not necessarily lead to less conflict; the fact that the number of land disputes is on the rise in Tanzania despite land law reforms over the past two decades hints at this sobering paradox.[53]

CONCLUSION

By mid-to-late 2016, the future of the EcoEnergy land deal hung by a thread. Several months following the prime minister's unexpected announcement in May 2016 that the government was shelving the project to conserve wildlife and their habitat in Saadani National Park, it was now rumored that the government was considering transferring the same land to another investor, again to establish a sugar plantation and factory. The rumor grew when the state media reported in October 2016 that President Magufuli had offered a ten-thousand-hectare land concession in an undisclosed location in the Coast Region, free of charge, to the family-owned Bakhresa Group, one of the largest domestic conglomerates, involved in a wide range of industries, including food and beverage manufacturing, transport and logistics, petroleum, and telecommunications.[1] According to news reports, the land in question became available after the president confiscated it from an investor "who had failed to use it for the benefit of the country."[2] Declaring this land grant at the inauguration of Bakhresa's fruit processing plant south of Dar es Salaam, the president praised Said Salim Bakhresa, the founding chairman of the group and one of the richest men in Tanzania, for being "the best example of businesspeople" and "the best taxpayer" in the nation, fit to lead the import substitution industrialization of sugar.[3] Only when Bakhresa established its newest subsidiary, Bagamoyo Sugar Limited, later that year did it become apparent that its ten-thousand-hectare plot would be superimposed on the very concession the government had granted EcoEnergy several years prior.[4] Despite being half the size, Bakhresa's parcel would encompass the most fertile, ecologically diverse, culturally significant, and historically contested floodplains on the south side of the Wami River, as illustrated in chapter 2.[5]

In June 2016, in light of the growing rumors about the project cancellation, I had asked EcoEnergy's communication officer, Saidi, about his views. We met at the company office in Bagamoyo town one Monday morning. The office had been sitting empty, with bare minimum furnishings collecting dust. After making a light comment about how dusty the place was and how their cleaner must have disappeared, he played for me a video recording of the prime minister's announcement from a month prior, which I had already seen. He stopped the recording midway and said, "This is totally contradictory to previous government efforts. We are fighting them every day."[6] He denied all rumors about the government confiscating EcoEnergy's land; the company had not received any official communication from the central government at that point. But he was clearly frustrated with what he saw as the lack of transparency and consistency in government actions:

> The government needs to have a real good reason to revoke this project.... The government has been the one that has been dragging their feet. They are a total chaos! We've already spent and lost over fifty million dollars waiting around for things to happen. The government wants to boost the economy and create jobs, and here you have a project that could employ potentially twenty thousand people. Instead of supporting it, the government is undermining it. We can't and we won't just give up. We can't pack up and leave. We will probably take the state to court.[7]

Finally, on November 9, 2016, the government notified EcoEnergy that its land title had indeed been revoked and the land returned to the president.[8] The following year, just as Saidi had predicted, the company took legal recourse, though not through litigation in court, but through a private means of dispute resolution. On September 11, 2017, EcoEnergy submitted a request for arbitration at the World Bank's International Centre for Settlement of Investment Disputes (ICSID), citing that the United Republic of Tanzania had violated its bilateral investment treaty with the Kingdom of Sweden.[9] I first learned about the arbitration from an email I received from EcoEnergy's executive chairman a few days after the case's filing. In his message, he regretted having "missed the opportunity to influence the agricultural development in Tanzania by doing a role model project" and stated that the company had no choice but to file an arbitration claim given the government's lack of interest in any dialogue. "It is not only a matter of principles and money," he wrote, "but primarily we owe it to the people of Bagamoyo and Tanzania."[10]

Represented by the largest law firm in the Nordic region, Mannheimer Swartling, EcoEnergy claimed that the company suffered a loss of USD 52 million plus future profits, as well as over ten years of work spent on project development, as a result of acts and omissions of the Tanzanian government.[11] The company alleged that Tanzania's failure to observe its investment treaty obligations resulted in a "lost opportunity to create approx. 20,000 new rural jobs affecting more than 100,000 people in Bagamoyo district."[12] The state's unilateral termination of the land deal, the company argued in the final paragraph of its 2017 white paper, would have far-reaching consequences for the nation: "The most serious long-term damaging effect is the loss of credibility for Tanzania as a country for much needed direct investments into agriculture. If the current GoT [Government of Tanzania's] attitude toward private investments and the rule of law is not altered, the ability to attract investments for modernizing and industrializing the Tanzanian economy risks being seriously undermined."[13]

Five years into the proceedings, an ICSID tribunal rendered its decision, or "award" as it is known, to both parties on April 13, 2022. While ICSID has yet to publish the details of the decision, Tanzania's attorney general confirmed with the *Citizen* that EcoEnergy had won a USD 165 million award against Tanzania.[14] Two months following the decision, Tanzania filed an application for annulment, the proceedings for which are pending as this book goes to press in April 2023.[15]

From its inception in 2005 to the ongoing arbitration in 2023, there was nothing definitive or inevitable about the land deal's trajectory. As historian Sara Berry observed, the conditions and outcomes of capitalist agrarian transformation in Africa have always been open-ended, shaped by a series of "inconclusive encounters" between the state and those it sought to control or enact rural development with.[16] Despite the termination of the land lease contract between EcoEnergy and Tanzania, the project has not entirely ceased to exist. It lives on, in varied forms, in the international legal sphere behind closed arbitral chambers, in national political debates and imaginaries in Tanzania, and in the bitter landscape and memories in Bagamoyo.

In this epilogue, I locate the unraveling of the EcoEnergy land deal post-2015 in the shifting political and economic conditions under the Magufuli regime, characterized by resource nationalism and authoritarianism. I then consider the unfinished legacies of both the EcoEnergy land deal and the Magufuli presidency not just for the "new" Bakhresa project but more importantly for people's lives and livelihoods. Drawing on the lessons of this book, I close with a critical reflection on the limits and contradictions of governing transnational land deals and the associated rise in investor-state arbitrations within the confines of the global investment regime.

Magufuli, Authoritarianism, and Socialist Nostalgia

The series of events I outlined above occurred at a particular conjuncture in Tanzania's history, during the presidency of John Pombe Magufuli, nicknamed "the Bulldozer," most notably for his uncompromising, hard-charging leadership style.[17] As discussed earlier in the book, Magufuli came to power in November 2015, replacing the Bagamoyo native and business-friendly Jakaya Kikwete, under whose backing the EcoEnergy land deal came into being.[18] Once in office, Magufuli wasted no time in following up on his campaign promise to crack down on government corruption and misspending. He famously paid surprise visits to government offices to ensure civil servants were doing their jobs; ordered all tax evaders to pay up or face prosecution; prohibited all but top government officials from flying first class; and slashed funds for lavish cocktail parties and dinner galas hosted by public institutions.[19] These bold gestures helped affirm his nickname and inspired among his followers in Tanzania and across the continent viral hashtags such as #WhatWouldMagufuliDo and #MakeTanzaniaGreatAgain.[20] In his first year in office, he also embarked on a nationwide campaign to unilaterally seize land left "idle" by so-called unproductive investors. The rationale for this campaign, according to the minister of lands William Lukuvi, was to redistribute land to citizens, poor farmers, and "those in need," although the government never specified how this putatively progressive agenda would be achieved.[21]

While Kikwete had enacted some protectionist measures during his second term, Magufuli's arrival heralded a marked shift in the way the state regulated natural resources on which the economy depended.[22] Most notably between 2017 and 2018, Tanzania passed new legislation that banned the export of raw materials for beneficiation outside the national territory and gave the president the permanent sovereign right to cancel contracts in the energy and resources sector or to remove "unconscionable" contract terms that jeopardized Tanzania's national interests. The new laws also required government shareholding in all mining and extractive projects and, following a slew of international arbitration claims by EcoEnergy and other companies, prohibited foreign investors from resorting to international dispute resolution mechanisms.[23]

Though careful not to dismiss the need for investments altogether, President Magufuli on many occasions stressed that the nation was fighting "economic warfare" against foreign companies that were "stealing from Africa."[24] Speaking of the largest gold miner in the country, Acacia Mining (formerly African Barrick Gold), which allegedly owed the government USD 190 billion in unpaid taxes, penalties, and interest accumulated over a period of seventeen years, Magufuli, for example, stated in 2017, "We need investors, but on a win-win situation and not those that exploit us. . . . They can come from the North, South, central and from anywhere but we must share in the profit. . . . Enough is enough! We have been given raw deals for too long and this has to end. . . . We cannot allow ourselves to be exploited forever."[25] The same year, in his address to Parliament, the minister for justice and constitutional affairs reiterated the president's call to fight against foreign corporate interests:

> We are not naïve; we are fighting giants. . . . But we are not afraid of giants! Let us try! There will be sabotage, definitely. But this country has survived sabotage, many years, and we must try [to fight] than to just surrender. . . . We cannot have investors with whom we do not benefit. . . . Let us not forget that this country has survived even greater threats than we are facing today. In 1964, if Mwalimu [teacher, a term of endearment for Julius Nyerere] had caved in to the Germans who were throwing a lot of money at us, if he had listened to them, we wouldn't have become a United Republic. . . . This nation has overcome difficult times, because we had leadership that had the courage to defend the interests of the people. This is another transition period.[26]

These statements are redolent of the socialist rhetoric of the 1960s and 1970s, in which the state not only promoted self-reliance as a nation-building strategy but also drew on the metaphor of bloodsuckers (*wanyonyaji*) to describe the economic exploitation of Africans by non-Africans, namely wealthy Europeans, Indians/Asians, and Arabs.[27] If Tanzania's turn to neoliberalism since the 1980s entailed a certain degree of "organized forgetting" on the part of state elites to dissociate themselves from the socialist past, the Magufuli regime called for a partial reversal of this trend: a conscious state-led effort to revive the nationalist ideology and the populist discourse of the ujamaa era to confront the ills of the global neoliberal economic order.[28] While xenophobic sentiments toward rich domestic racial minorities, particularly Asian- and Arab-Tanzanian business elites, have relatively weakened, as exemplified by Magufuli's preferential treatment of Bakhresa, Magufuli and the ruling party, CCM, increasingly adopted the view that Tanzania has been mired in a "rot" because the nation strayed from the developmental vision of the nation's founding father.[29] On Nyerere Day 2018, a public holiday to commemorate Nyerere's death, Magufuli vowed to follow in the founder's footsteps and urged Tanzanians to remember his legacy: "If you read the Arusha Declaration, you know everything is in there, everything that our *nation was supposed to build* is in there."[30]

Magufuli's resource nationalism, which he justified with nostalgic appeals to "Nyerere's Tanzania," went hand in glove with his authoritarian grip on society. Increasingly throughout his presidency, Magufuli clamped down on freedom of speech and peaceful assembly; prohibited opposition parties from holding rallies and protests; banned pregnant girls and young mothers from attending public schools; and harassed, abducted, and arrested dissenting politicians, activists, journalists, businesspeople, artists, and ordinary citizens alike for alleged sedition.[31] Tundu Lissu, his staunchest political opponent, who was nearly assassinated in September 2017, likened Magufuli to a "vicious tyrant" who sanctioned "a Mafia-style shakedown on a gigantic scale."[32]

At the outset of the global pandemic in 2020, Magufuli also dismissed the head of the national health laboratory and stopped releasing data on the number of COVID-19 cases. Instead, he asked all citizens to pray to chase the "satanic" virus away, and in June 2020 declared that the nation had rid itself of the disease, "thanks to God."[33] On the eve of the national election later that year, the UN high commissioner for human rights, Michelle Bachelet, warned of "a deeply deteriorated environment for human rights" in Tanzania.[34] The government, meanwhile, denied accreditation to foreign media and blocked local and international missions from observing the election. For days before and after the election day, many Tanzanians reported sudden disruptions to the internet and social media platforms, which critics argued was a CCM-orchestrated effort to suppress opposition in the face of a real possibility of the party's first electoral defeat since independence.[35]

On November 5, 2020, Magufuli was sworn in for a second term amid allegations of election fraud.[36] The National Electoral Commission, whose director is appointed by the president, declared that Magufuli had won a landslide victory with 84 percent of the popular vote.[37] The election also resulted in the ruling party claiming a whopping 97 percent of 264 elected parliamentary seats.[38] Opposition leaders who called for mass protests against the election results—including Chadema's presidential candidate, Lissu—were swiftly detained, charged with "terrorism" and the organization of "unlawful" assembly.[39] The tightening of security by the police and army ahead of the swearing-in ceremony effectively deterred any potential demonstrations.

In an ironic turn of history, Magufuli died on March 17, 2021, at the age of sixty-one, following a mysterious two-week disappearance from the public scene.[40] Though Vice President Samia Suluhu Hassan, who subsequently became the nation's first female and Zanzibari president, attributed his death to chronic atrial fibrillation, his opponents contend that the actual cause was COVID-19 complications.[41]

In the first year of her "accidental" presidency, Hassan, or "Mama Samia" as she is popularly known, has embarked on several changes to reverse her predecessor's policies. She has reshuffled the cabinet twice, bringing in prominent figures from the Kikwete era and expelling those perceived to be Magufuli loyalists.[42]

To rebuild Tanzania's international reputation, she created a new scientific task force to address the pandemic and pledged to restore democratic rule of law. She also promised to offer incentives to foreign investors, particularly in the agriculture, energy, and conservation/tourism sectors, all of which demand pressures on rural landscapes and livelihoods.[43] As the new administration increasingly re-embraces neoliberal ideals, however, the resource nationalist measures Magufuli put in place still remain in effect.

"Tanzania only problematically qualifies as a 'postsocialist' state," argued anthropologist Kelly Askew upon observing that Tanzania has yet to officially denounce socialism and discard socialist elements that are enshrined in national laws, particularly with regard to land and natural resources.[44] If two primary features of socialism are the social or national ownership of the means of production and the relative monopoly of political activity by one party, the recent developments in Tanzania appear to reinforce this argument. Though premature and unexpected, Magufuli's death does not negate or lessen the significance of the changes he set in motion during his tenure, including the reappropriation and reallocation of the land in Bagamoyo to the Bakhresa Group and the repression of dissenting voices, including those on the precipice of displacement and dispossession. In what follows, I offer a glimpse into how the ongoing legacies of the Magufuli regime and the EcoEnergy land deal have transformed and continue to reverberate in Bagamoyo's landscape and the lives of ordinary rural citizens, with profound gendered and generational consequences.

The Remaking of a Bitter Landscape

The trajectory of Bakhresa's Bagamoyo Sugar Limited has been as inconclusive as Tanzania's postsocialism and its partnership with EcoEnergy.[45] In January 2018, an article in the *Citizen*, titled "What Holds Up Sh660bn Bagamoyo Sugar Project," hinted at initial signs of project delay.[46] Bakhresa's corporate affairs director had told the reporter then that they were waiting for the Ministry of Lands to complete the valuation report and compensate the local people who would lose their land for the project.[47] This of course was not a new problem but one that had confounded every round of enclosure in Bagamoyo since the nation's founding, as chapter 2 demonstrated.

Notwithstanding the initial setback, the government moved much more swiftly with the compensation process than it had done during its years-long partnership with EcoEnergy. By April 2018, the central government had ordered regional and district authorities to instruct local residents to start preparing for their move. Overcome by a sense of foreboding, twenty elders from Matipwili village, all of them floodplain farmers who have experienced multiple threats of eviction and actual displacement in their lifetime, wrote a letter to the regional commissioner, seeking answers about their future. The letter, drafted by a male elder named Suba on behalf of his fellows, implored,

> When we are evicted from our homes and farms, where should we stay? Where should we farm and how should we live? How should we educate our children? Farming means everything in our lives.... We should be given housing and land, and only then we shall leave. A review of compensation should be conducted to ensure justice. Information about payment is not clear and no one knows what they will receive.... Finally, we ask the Regional Commissioner to have compassion and to try to see things from our perspective.... We wish you good work in Nation Building [*Ujenzi wa Taifa* (sic)].[48]

Despite the elders' poignant plea and allusion to Nyerere's unfinished socialist nation-building project in which rural farmers once figured centrally and which Magufuli sought to revive, evictions began four months later, in August 2018. In a news segment aired August 16 on Azam TV, Bakhresa's subsidiary media company, the coast regional commissioner is shown visiting the project site, ordering all current occupants or "intruders" to leave the land immediately.[49] He says, "The president of the United Republic of Tanzania is a very good person, and he has already said that no human rights will be lost.... But also, the same president forewarns all people who think they can get free money from the government. That is impossible; [only] those who are entitled will get compensation."[50] To determine eligibility for compensation, the government used the data from the PPC conducted for the EcoEnergy Sugar Project in 2011 and 2014 (see chapter 3), rather than starting the process anew. The twin imperatives of time and money arguably triumphed over the fact that people's lives and social relations on the land have changed over the many years since the initial valuation. As the government's chief valuer had told me in an interview in 2016, "There are budget constraints to valuation. There is no revaluation in any case; [we] only update the payment amounts based on differences in annual crop values and interest rates."[51] In privileging bureaucratic convenience, the state effectively rendered static and immutable the ever-changing agrarian social relations, while strengthening its authority in the countryside.[52]

Displaced residents I was able to track down lamented how there was little room for dissent, debate, or dialogue about the compensation process, reflecting both the diminishing democratic space under Magufuli rule and a pattern repeatedly observed in studies of compulsory land acquisition and compensation in Tanzania.[53] Whereas some grassroots resistance efforts were simply ignored, like the letter the Matipwili elders sent to the regional commissioner, other forms of direct action were promptly quelled by the police. Rama, the wanaharakati secretary introduced in chapter 5, described one incident:

> There was one time when the prime minister came to visit the [Bakhresa] project site. We gathered on the side of the road to make noise, to make sure he heard our cries as his car passed, but the police quickly came and forced us to scatter. They beat us with rods and told us we should not be seen here. They said if we don't scatter now, they are going to catch us and throw us in jail. Frankly, many people got scared and fled after that. I wanted to continue protesting, but I was all by myself.[54]

Suba, the Matipwili elder, similarly relayed that under the threat of violence and the exigencies of displacement, no semblance of collective action could be sustained.[55] "We all wanted to protest," he said, "but officials from the district land office told us that if we resisted, we would be denied any sort of compensation, and there would be consequences like being arrested. We could lose not only our land, our fields, our homes, but also whatever money that was set aside for compensation. There was verbal abuse, and we felt forced to accept it as our fate."[56]

When it came to compensation payments, inconsistencies and exclusions were rampant, revealing and exacerbating multiple intersecting inequalities. To claim compensation, people first had to submit to the district land office the original forms they received during the PPC: the Land Form 69a and Valuation Form 1. These forms, as discussed in chapter 3, assumed the nuclear family as the basic unit of society and privileged husbands, fathers, and male heads of households as primary beneficiaries of compensation. If people had lost those forms, had only photocopies, had one form but not the other, or were not named on the forms, the government refused them payment. Claimants were also required to have bank accounts to which compensation could be deposited; only when bank details were provided were the payment amounts disclosed. This created a significant barrier for the poorest residents, including the few women who had become eligible for compensation by virtue of their unmarried status, who could not afford to travel to town to visit banks, let alone have the official national identification cards required to open bank accounts.[57]

Based on information I could gather from twenty displaced families, the payments ranged drastically from none to TZS 40 million (USD 20,000), with an average of TZS 4.5 million (USD 2,250) and a median of TZS 500,000 (USD 250) per household. Bambadi allegedly received the largest payout in the sample, although he had originally rejected the PPC and a High Court verdict had rendered him a trespasser, as the previous chapter noted. Migrant residents, by and large, were excluded from government compensation, but some reported that a male Bakhresa project manager offered them a "gift" of TZS 100,000 (USD 50) per person, saying that the company wanted to remain "good neighbors" with its potential future employees.[58] The government also denied compensation to pastoralists, because they could not demonstrate proof of landownership, permanent residence, or crop cultivation, a rationale that has historically justified state oppression of ethnic-minority nomadic peoples since colonial times.[59] In the case of elders, several had passed away before they could even claim compensation.[60]

Among those who received compensation, confusion abounded on what the payment precisely included or how it was calculated. As Rama described,

> The payout was done carelessly. It was chaotic. For example, on the farm you might have had trees, bananas, mangoes, and so on. On the form you might have had those trees listed, but the maturation rates were all wrong because the valuation was done many years ago. So you go to the district land office and ask, "Does the amount I received reflect what I currently have on my land?" The officials would tell you nothing, they would show you nothing. They would ignore you or just say, "This is the fare that will help you get to wherever you need to go. So just take it and leave."[61]

For Mwajuma, an elderly widow and floodplain farmer from Matipwili whose photo-narratives appeared in chapter 2, no amount of compensation could make up for the loss of land. Losing land, as she described it, meant losing all her relations, immeasurable pain from which there was no moving on or letting go:

> Many people opposed compensation. We didn't want to abandon our land. But the government told us if we had agreed to be counted back then [for the EcoEnergy PPC], then we had already agreed to have our land taken for development purposes. We were scorned by district officials. They said they didn't need our advice on how to calculate compensation. But who could know better about the value of our land than us? We have been the guardians of the land all this time. How do you put a price tag on the land you built your entire life on? How is that possible? The land was everything I had. I knew it by heart. It was like family. The land was like my child, I took care of it, and it cared for me. It's like the investors have kidnapped my child and I don't know when I will be able to see them again. It pains me every day that I lost my land. I long to return to my land. My soul will not rest easy until I get my land back.[62]

The absence of formal grievance mechanisms intensified the state of unrest and anguish people felt, as it deepened gendered inequalities at the same time. For example, Mohammedi, a bar owner and a relatively well-to-do male farmer introduced in chapter 5 who received TZS 6 million (USD 3,000) in compensation, said he complained personally to the district executive director and managed to secure an additional two million shillings in cash from the government. On the other hand, Mwanahamisi, Mwajuma's younger sister and a married floodplain farmer whose photo-narratives also appeared in chapter 2, said she tried in vain to file a complaint to the district land office about her husband, who refused to share the TZS 13 million (USD 6,500) he reportedly collected on behalf of his family. With no offer of assistance from district authorities and after much bickering and bargaining with her husband, she managed to persuade him to give her TZS 2.5 million (USD 1,250) as her share, but no more. According to other married women I spoke with, what Mwanahamisi was able to achieve was exceptional; most refrained from complaining because they worried that their husbands would retaliate in violence or run away with the money.[63]

Whatever grievance people might have had, they were forced to vacate the land in a matter of days upon receiving payment. As Rama recounted,

> The district commissioner told people that we had three days to leave the project area after receiving our payments. We had many questions, but he ignored us and said he was simply carrying out an executive order from the president. We didn't know what to do, where to go, so we decided to wait and see. Three days later, the commissioner said he would give us four more days, giving us seven days in total. On the eighth day, we were shocked to find out that Magufuli had sent a bulldozer without prior warning. He sent it to demolish our homes.[64]

Previously a member of Chadema, Rama said he decided to defect to CCM after the wanaharakati dissolved, hoping to seek favors from the ruling elites. His new loyalty to the party helped spare his house from being demolished, but only briefly:

> A Bakhresa project manager called me on the day of the demolition. I was away at the time, trying to find new land to farm. The manager called me to come back home immediately. He said, "All houses here have been demolished except for yours. We respect your house because it has a CCM flag in front of it." When I arrived, everything I had inside my house had been thrown out. There was only one thing that was left inside: an envelope with CCM cards [for registering new party members]. There was a reporter from Channel 10 there. When I came out of my house, I deliberately dropped the cards, and that's when the reporter started photographing me. The project manager quickly covered the camera and said, "No, don't take pictures, because if you do, then you

will be offending the CCM national chairman, Mr. President, and if he sees the photograph, it will be like we are insulting him."

As in the past, the mobilization of state violence was indispensable to enacting dispossession. Rama recalled the day in despair:

> I tried to stop the people that came for the demolition, but they were intimidating. There was a great army, soldiers with guns; one might have thought they were going to war. They were in military uniform. But some of them didn't look like they were real soldiers; some even had dreadlocks, and that is not the way of our men. The soldiers said they were sent by the district commissioner, who said we were all just vagrants here. It pained me so much to see them destroy my house right in front of my eyes.
>
> To this day, we fail to understand how the government could be so cruel to us, how we got to a point where we are building houses with twigs and tarps. There are so many children who are not going to school. We do not know our fate. Speaking for myself, I also do not know my fate. Having no permanent home means I cannot settle down here forever. A man needs his land, farm, and house to provide for his family. I have lost everything I have invested in. I have fallen behind.[65]

Since being evicted, most migrant residents and those of younger generations like Rama have moved to places like Bozi and Kibuyu Mimba to find new farmland or to wait for possible job opportunities that might come up in the near future at the sugar plantation and factory. Some have moved closer to Bagamoyo town to take up odd jobs there, while others left Bagamoyo permanently in search of opportunities elsewhere. Pastoralists have shifted their *boma* (homestead) to the western margins of the railway tracks, but they continue to rely on the project site and other cultivated areas nearby to graze and water their livestock, at the risk of being fined or igniting sometimes violent farmer-herder conflicts.[66]

Long-term residents and floodplain farmers, many of them elderly, on the other hand, have moved north across the river and resettled in other subvillages of Matipwili. According to Suba, most of these farmers, including himself, have not been able to resume their livelihoods:

> There are so many challenges. Cutting down trees is the first step in preparing a new farm. It's a laborious task; even young men have a hard time with it. You also feel bad clearing the forest and uprooting young trees because you know they help reduce soil erosion, but what choice do you have? Preparing the farm, planting, waiting for harvests, and learning the habits of the soil all take time, and so when we shifted to

this side of the village [north of the river, near the village center], we had nothing to eat. Where we farm now, there is not enough water. We are not as close to the river as we were before, and so that means no fishing either. Because of these challenges, our economic situation has gotten worse. Hunger is a problem for elders and widows who stopped farming. For old people like me, it feels like I am just waiting for the end of my life on Earth.[67]

While floodplain farmers used to rely on their neighbors (many of them siblings and extended family members) during lean times, the breakdown of this support system following displacement severely impaired their possibilities for social reproduction. When I last spoke with Mwajuma in October 2020, she had given up farming and had been trying to make ends meet, working as a casual farm laborer: "I don't have the desire to clear new land now, because I am afraid it will be grabbed again. At this old age, it is not worth the effort. And we are too far from the river. I have been asking larger farmers around here if they need help with plowing, weeding, etc., so I don't starve."[68]

The testimonies above, necessarily partial but nonetheless important and heartbreaking, illuminate the complex material, symbolic, and affective registers of displacement and dispossession. The expressions of injustice and loss—the wretched straits of building homes with "twigs and tarps," the belief that one has "fallen behind" on one's life and economic goals and societal gender expectations, the inability to "rest easy" until land is returned, and the feeling that one is "waiting for the end of their life on Earth"—all index a renewed and deepening sense of precarity and liminality of agrarian life. The uneven and unsettling process of becoming uprooted reproduced existing inequalities and forced complex renegotiations of one's identities and subjectivities.

While Bakhresa has seemingly been able to move further along than EcoEnergy in its project implementation, nothing is determined about its trajectory. Although the government and the company may appear to have "fixed" the problem of population control through unjust removal, the government has yet to address another controversial issue that remains unresolved: the boundary dispute regarding Saadani National Park. This very issue reemerged in public debate around the same time the ICSID tribunal rendered its award in the dispute between EcoEnergy and Tanzania. In an annual report the National Audit Office released in April 2022, the controller and auditor general warned the president of the "significant impact" the Bagamoyo sugar project would have on environmental and wildlife conservation. He noted that over a third of Bakhresa's land concession was located within the national park, and that activities of the sugar plantation and factory, including nutrient and chemical runoff, would pollute the wetland and coastal ecosystem and endanger wildlife health and reproduction.[69] The report made no mention of the impacts, both potential and already experienced, on human populations, a sharp reminder of the continuing neglect and marginalization of rural lives. In summary, many questions remain regarding the future of the Bakhresa land deal: how commercial production will unfold; to what extent and in what gendered ways the plantation will incorporate or exclude the dispossessed populations; how industrial monoculture will transform local ecologies and societies; and what kinds of contentious politics and resistance will emerge and continue to reshape Bagamoyo's bitter landscape.[70]

Land Deals and Limits of the Global Investment Regime

EcoEnergy is the first known foreign investor directly implicated in the global land rush in Africa to have filed an arbitration claim against the host government. Since the case's filing in 2017, Tanzania has become a respondent to five other international arbitration claims, four of which pertain to foreign land-based investments in the agricultural, mining, and energy sectors. All of these cases have arisen in connection with the Tanzanian government's alleged seizure of the investors' land, the unilateral termination of contracts, or both.[71] These foreign entities could file arbitration claims against the government by invoking the investor-state dispute settlement clause, which has become a standard feature of international investment agreements, including the bilateral investment treaty Tanzania and Sweden signed in 1999.[72]

The use of international arbitration, as an alternative, nonjudiciary dispute resolution mechanism gained momentum in the 1990s with the expansion of neoliberal economic governance. Since then, the number of cases has grown exponentially, with US and European companies driving much of the global arbitration boom.[73] In the early days of the global land rush, observers had already anticipated a rise in arbitration cases associated with cross-border land deals. "Given the haste and lack of forethought going into some of these deals," one publication warned in 2009, "it seems all too likely that they will give rise to a great deal of arbitration in the years or even decades to come."[74]

The dispersal of land deal politics to the international legal sphere, specifically ad hoc arbitral tribunals, raises several causes for concern. Who gains and who loses from international arbitration cases linked to canceled land deals? What implications do these cases have, regardless of their outcome, for rural communities such as those in Bagamoyo? First, it is important to understand that the costs of arbitration proceedings weigh heavily on respondent governments in large legal fees and, in cases where they are found responsible for violating treaty obligations, in millions of dollars in compensation payments to investors. Tanzania currently owes at least USD 185 million for a lost arbitration against the Hong Kong–based Standard Chartered Bank involving a dispute over rights under a power purchase agreement,[75] and the minimum compensation claim in a recent arbitration that a UK-based mining company initiated in 2020 stands at USD 95 million.[76] These costs are bound to have serious consequences for governments in the Global South, adding significant burdens to already constrained national budgets and existing debt service obligations.[77] More critically, they could divert public spending away from much-needed investments in areas such as agrarian reform, land redistribution, and social protection that have the potential to reduce poverty and inequality, advance gender justice, and improve the living standards of millions of rural producers and workers.

Second, as many critics have argued, the global investment regime is fundamentally biased toward corporate interests. Whereas aggrieved foreign investors can initiate arbitration proceedings against their host governments, those governments and the communities impacted by investors' activities cannot file their complaints in the same tribunals.[78] An exemplary case in point: the hundreds of rural women and men whose lives the EcoEnergy project upended are completely unaware of and uninvolved in the ICSID arbitration. This flies in the face of EcoEnergy's contention that they "owed" the arbitration to "the people of Bagamoyo and Tanzania." Although states have the moral obligation to represent the interests of their citizens, they may not be willing or able to effectively argue before the tribunal from the perspectives of local populations.[79] Since investor-state disputes are adjudicated primarily on the basis of investment treaties, their limited scope further prevents consideration of public interest, including the kinds of injustices and rights violations the foregoing chapters examined. International arbitration proceedings also lack transparency by being closed to the public, and it is often only a handful of private law firms, elite arbitrators, and financial consultants based in key arbitration hubs such as Washington, New York, Paris, London, and Singapore that reap financial benefits from the current global investment regime.[80] Ultimately, international arbitration—along with the network of actors, processes, and relationships that sustain it—is both constituted by and constitutive of global capitalism. It is a market-based mechanism that disembeds contemporary land deals from society and abstracts them from the lived and contested landscapes with which they are co-produced.[81]

Third, the protections investors derive from existing investment treaties can fuel more land grabbing and undermine ongoing struggles for land and food sovereignty by local communities and transnational social movements.[82] Although international development organizations and financial institutions have promoted various "best practice" standards for governing land-based investments over the past ten years—including the Voluntary Guidelines on the Responsible Governance of Tenure of Land, Fisheries, and Forests in the Context of National Food Security; the Principles for Responsible Investments in Agriculture and Food Systems; and the Guiding Principles on Large-Scale Land Investments in Africa—they remain nonbinding and their implementation woefully limited. This begs the question of why land investments should be governed through voluntary means, while existing investment treaties between investors and states are legally binding.[83] Perhaps the most problematic aspect of voluntary standards is that they assume large-scale, capital- and energy-intensive agricultural and extractive land deals as necessary and inevitable.[84] By defining land and investment in narrow, economically reductionist terms, current global governance frameworks limit the possibility of imagining and enacting radically different, anticapitalist ways of organizing society and the environment. As Olivier de Schutter, the former UN special rapporteur on the right to food, has sharply criticized, these existing guidelines provide nothing more than "a checklist of how to destroy the global peasantry responsibly."[85] As long as investment treaties remain in force, they give foreign investors the license to file arbitration claims against host governments, independent of whether they themselves have acted in accordance with existing voluntary codes of conduct and national laws. And as long as existing investment guidelines remain voluntary, and unless such guidelines are able to challenge the fundamental power imbalances that underpin the liberal international investment regime, they will continue to shield both investors and states from scrutiny of accountability and justice.

"To be rooted is perhaps the most important and least recognized need of the human soul," wrote Simone Weil.[86] Being uprooted and estranged from the land, both in situ and ex situ, over a period of a decade (and counting) as a result of the new enclosures inflicted material, emotional, and psychological wounds on the people of Bagamoyo. The resulting scars, which neither the state nor the investor has yet to recognize and take accountability for, will be long inscribed and sedimented in the landscape and felt for generations to come. In closing this book, however, I choose not to end my analysis simply on a note of damage and despair. As decolonial feminist scholar Eve Tuck writes, to re-envision our theory of change, researchers must document "not only the painful elements of social realities but also the wisdom and hope."[87] Desire, above all, is about longing—longing for change that is conditioned and enriched by the lessons from the past, the politics of the present, and possibilities for the future. This kind of yearning, bell hooks writes, "wells in the hearts and minds" of those who have been silenced and marginalized by grand narratives of progress and modernization.[88]

What people hoped for in life in Bagamoyo, as I have come to learn through years of sustained ethnographic engagement, was simple but no less significant. They wanted to live on and with the land, plant trees, build homes, and raise their children in safe and nurturing environments. I am reminded of Daudi from chapter 5, who, before his passing, expressed his desire to farm without fear of violence and do mundane but meaningful things like planting orange and coconut trees. Mwanahamisi, too, before her eviction, spoke of her intention to continue the fight for land: "Our land here in the coast does not die. Like our land, our struggle here will not die. Our fight for land is not just for us, but for our children and grandchildren who will grow up to use this land. That's why we plant trees, to mark our place, our roots, and to remember how the air smells when the flowers bloom and the fruits ripen."[89]

As this book has highlighted, acts of resistance, reoccupation, and repossession have historically accompanied each and every round of enclosure and dispossession in Bagamoyo, however fragmented, gradual, and impermanent these efforts might have been. This insistence on presence and subsistence is what has given Bagamoyo's landscape and people a fighting spirit. No enclosures are ever complete.

Glossary of Swahili Terms

This is a limited glossary of Swahili terms. Only words that appear more than once or hold ethnographic significance are included in the list.

ardhi land
banda shed, hut
chachu yeast; a thing or action that provokes
gongo strong distilled liquor
kanga colorful wax print fabric with a central design motif and patterned border that includes Swahili proverbs or aphorisms
kiangazi dry season
kilimo agriculture (from *kulima:* to farm, cultivate)
kitenge (pl. *vitenge*) colorful wax print fabric containing a variety of patterns that is thicker than kanga
kitopeni an area that becomes muddy or wet after the annual inundation of the floodplains; colloquially, a muddy season or the period between the long and short rainy seasons in the coastal riverine floodplains
kungwi (pl. *makungwi*) female initiation instructor (from *kunga:* to teach secret knowledge)
masika long rainy season
maulidi Islamic religious celebration
mgambo paramilitary, guards, auxiliary police (short for Jeshi la Mgambo, People's Militia)
mkole false brandy bush (*Grewia bicolor*), a species of shrub with flexible branches and small edible fruits; a vernacular term that signifies fertility and female initiation rites for matrilineal coastal ethnic groups, such as the Zaramo, Kwere, Zigua, and Luguru
mvumo deleb palm (*Borassus aethiopum*), a species of African palm with robust trunks and fan-shaped leaves
mwali female initiand
mwenyeji (pl. *weyeji*) local person, native to a place
mzee (pl. *wazee*) old person, elder, or a respectful term for husband
mzungu European or Western person; foreigner with connotation of whiteness
ndago yellow nutsedge (*Cyperus esculentus* L.), a species of reedlike grass with resistant roots that grows in wet areas along the rivers
ngoma drum and dance celebration
serikali government
taifa nation
ujamaa familyhood; a version of African socialism led by Tanzania's first president, Julius Nyerere
wanaharakati activists
watu people

Notes

INTRODUCTION

1. Songa, "Bagamoyo Embarks."
2. Songa, "Bagamoyo Embarks."
3. As I will elaborate in chapter 1, the agreement was that the government would initially acquire 10 percent equity shares in the project company upon financial closure, or when all financing agreements with lenders and the required conditions contained in them had been met. These shares would rise up to 25 percent after eighteen years of commercial operation of the project. All equity shares would be nondilutable, meaning the government would maintain its 25 percent level of ownership should the company issue additional shares in the future.
4. Elgström, "Giving Aid."
5. Agro EcoEnergy Tanzania, "Early Works Launch."
6. Kilombero Sugar Company, established in 1961, is the largest sugar producer in Tanzania. It is majority owned by Africa's largest sugar producer, the South Africa–based Illovo Sugar. Illovo Sugar is a subsidiary of Associated British Foods. Previously state-owned, the Kilombero sugar estate became privatized in the late 1990s in the wake of structural adjustment programs. Though Kilombero was presented to Bagamoyo farmers as a successful example of an outgrower scheme, researchers have argued that it has historically benefited larger outgrowers at the expense of small producers and exacerbated existing class and gender inequalities. See Sulle, "Social Differentiation"; Sulle and Dancer, "Gender, Politics, and Sugarcane." On the contradictions of contract farming schemes in sub-Saharan Africa see also Little and Watts, *Living under Contract*.
7. AfDB, "Bagamoyo Sugar Project." An earlier environmental and social impact assessment for the project published in 2008 reported that there were approximately six thousand people residing in the villages and subvillages "bordering" the planned project site.
8. The project agreed to reserve two thousand hectares of the concession area and negotiate an additional twenty-five hundred hectares or so with neighboring villages for pastoralists (Agro EcoEnergy Tanzania, "Draft Resettlement Action Plan"). Few farmers and pastoralists I interviewed were aware of these details, but for those who knew, it was a point of bitter contention. According to a young pastoralist man, these promises of land were not enough, given the number of livestock each family had, as many as one thousand cattle; for farmers, they found it preposterous that they would be evicted while the pastoralists, whom they perceived as "outsiders," were rewarded with land (field notes, various dates, 2013–2016). Most pastoralists in the concession area belonged to the Datoga-speaker Barabaig, one of the most marginalized ethnic minority groups in Tanzania. Many of them had been displaced, or were descendants of those who had been displaced, from their ancestral homeland in the Hanang Plains of north central Tanzania in the 1970s by a large-scale wheat mechanization scheme sponsored by the Canadian government. While I was able to interview several Barabaig households through a Datoga-speaking interpreter, the language barrier prevented me from sustained engagements with them. I recognize this as an important shortcoming of my ethnography.
9. Government land valuer, interview with the author, Bagamoyo town, July 31, 2014.
10. Songa, "Sh1 Trillion Sugar Deal Vanishes."

11. *Citizen*, "Project Axed to Save Wildlife."
12. Makoye, "Villagers Spared Eviction."
13. Field notes, various dates, 2015–2016.
14. This idea, of course, rests on Marx. As he wrote, "Men [sic] make their own history, but they do not make it as they please; they do not make it under self-selected circumstances, but under circumstances existing already, given and transmitted from the past." See Marx, "Eighteenth Brumaire," chap. 1.
15. On "in situ displacement" see Feldman and Geisler, "Land Expropriation and Displacement"; Feldman, Geisler, and Silbering, "Moving Targets." On "slow violence" see Nixon, *Slow Violence*.
16. Athumani, interview with the author, Bozi, Bagamoyo District, January 9, 2016. All names used in this book, except for elected officials, public figures, and those that appear in public record, are pseudonyms.
17. Neema, interview with the author, Bozi, Bagamoyo District, August 27, 2016.
18. Nuru, interview with the author, Matipwili, Bagamoyo District, September 24, 2016.
19. Williams, *Marxism and Literature*, 129.
20. See, for example, Mignolo, "DELINKING"; Quijano, "Coloniality of Power"; Mitchell, "Economics"; Connell, "Why Is Classical Theory Classical?"; Rodney, *How Europe Underdeveloped Africa*.
21. Kautsky, *Agrarian Question*; Lenin, *Development of Capitalism in Russia*. For a collection of essays that explores the key debates and contemporary relevance of the agrarian question see Akram-Lodhi and Kay, *Peasants and Globalization*.
22. Hart, "Development Critiques." See also Ferguson, *Anti-politics Machine*; McMichael, *Development and Social Change*; Cowen and Shenton, *Doctrines of Development*; Rist, *History of Development*; Mawdsley and Taggart, "Rethinking d/Development."
23. URT, Tanzania Development Vision 2025.
24. Outgrower specialist / EcoEnergy contractor, interview with the author, Bagamoyo town, August 19, 2013.
25. On the analytic significance of "the new enclosures" see Midnight Notes Collective, "Introduction to the New Enclosures"; White et al., "New Enclosures"; Geisler and Makki, "People, Power, and Land." On the coloniality and whiteness of development see Pailey, "De-centering the 'White Gaze.'"
26. Ferguson, *Expectations of Modernity*, 23.
27. Chung and Gagné, "Understanding Land Deals in Limbo."
28. Johansson et al., "Green and Blue Water Demand."
29. Land Matrix, "Dynamics Overview Charts." In addition to "intended" and "failed" land deals, the database, as of April 2022, reports 3,106 "concluded" land deals on approximately 149 million hectares. These numbers include transnational land deals for both agricultural and nonagricultural projects, such as conservation, tourism, mining, and oil and gas extraction. For a useful review of agricultural land deals reported in the Land Matrix database see Lay et al., "Taking Stock of the Global Land Rush."
30. Moyo, Yeros, and Jha, "Imperialism and Primitive Accumulation"; Carmody, *New Scramble for Africa*. For an important discussion on the methodologies and challenges thereof for studying the global land rush see Scoones et al., "Politics of Evidence"; and Locher and Sulle, "Challenges and Methodological Flaws."
31. Zheng, "China, Tanzania in Talks." In June 2021, President Samia Suluhu Hassan announced she would revive the Bagamoyo port project suspended by her predecessor two years prior; the negotiations between the government and the Chinese investor (China Merchant Holdings) remain pending as of this writing. See Ng'wanakilala, "Tanzania Says Resumes Talks."

32. Wise, "Picking up the Pieces"; Wise, "What Happened to the Biggest Land Grab"; Oakland Institute, "Understanding Land Deals"; Oakland Institute, "Losing the Serengeti"; Oakland Institute, "Looming Threat of Eviction"; Gardner, *Selling the Serengeti*. Despite temporary halts, delays, and cancellations, these deals, like the EcoEnergy case, did not result in the return of land to original users.

33. Chung and Gagné, "Understanding Land Deals in Limbo."

34. Wise, "What Happened to the Biggest Land Grab."

35. GRAIN, "Failed Farmland Deals." Other scholars have examined "failed land deals," including Schlimmer, "Caught in the Web of Bureaucracy?"; Gagné, "Resistance against Land Grabs in Senegal"; Schönweger and Messerli, "Land Acquisition, Investment, and Development."

36. Ferguson, *Anti-politics Machine*; Escobar, *Encountering Development*; Scott, *Seeing Like a State*; Schroeder, *Shady Practices*; Mosse, *Cultivating Development*; Li, *Will to Improve*.

37. Smith, "Women, Class and Family," 4.

38. Turner, "Variations on a Theme of Liminality," 68.

39. I emphasize spatiotemporality rather than simply temporality in my study to signal the ways spatial (e.g., land, water, the commons, places of dwelling, displacement) and temporal (e.g., the seasons, agricultural cycle, waiting, anticipation) relations and processes shape agrarian life. For excellent works that focus specifically on the politics of time or the temporality of waiting I would refer readers to Guyer, "Prophecy and the Near Future"; Auyero, *Patients of the State*; Masquelier, *Fada*; Gonçalves, "Orientações Superiores"; Fent and Kojola, "Political Ecologies of Time and Temporality"; Stasik, Hänsch, and Mains, "Temporalities of Waiting."

40. Anzaldúa, "(Un)Natural Bridges, (Un)Safe Spaces," 1; Anzaldúa, *Borderlands*.

41. See, for example, Bhabha, *Location of Culture*; Spivak, "Explanation and Culture."

42. Feminist geographer J. K. Gibson-Graham (*Postcapitalist Politics*, xxix) refers to reframing as a "technique of thinking" that "can create the fertile ontological ground for a politics of possibility, opening the field from which the unexpected can emerge." Reframing d/Development as always liminal, following no knowable paths and moving toward no predetermined ends, forces us to do exactly what Gillian Hart has urged critical scholars to do: "to confront questions of capitalist development—not as an unfolding teleology or immanent process, but in terms of the multiple, non-linear, interconnected *trajectories*." See Hart, "Development Critiques," 655.

43. Locher and Müller-Böker, "'Investors Are Good'"; Abdallah et al., "Large-Scale Land Acquisitions."

44. Here I draw on Anna Tsing's notion of friction, which she defines as "the awkward, unequal, unstable, and creative qualities of interconnection across difference." See Tsing, *Friction*, 4.

45. Sekab BT, "Environmental and Social Impact Statement."

46. Chung, "Engendering the New Enclosures."

47. See chapter 2.

48. Official in the Ministry of Lands, interview with the author, Dar es Salaam, August 20, 2014.

49. SAGCOT, "SAGCOT Investment Partnership Program," 14. The abbreviation "ha" stands for hectare.

50. Fairbairn, *Fields of Gold*, 26. As I describe in chapter 2, the nationalization of land in Tanzania is an artifact of colonial rule.

51. URT, Land Act, 1999. On compulsory acquisition for public purpose in Africa see also Alden Wily, "Compulsory Acquisition."

52. Chung, "Governing a Liminal Land Deal"; Chung and Gagné, "Understanding Land Deals in Limbo."
53. Ribot and Peluso, "Theory of Access," 154–55.
54. Elsewhere, I term this perspective "feminist ontology of land." See Chung, "Engendering the New Enclosures." Alice Beban expands on the concept to highlight land as "not only life producing but life itself." See Beban, *Unwritten Rule*, 206.
55. Many works have inspired my thinking on landscapes. Some of these include Mitchell, *Landscape and Power*; Neumann, *Imposing Wilderness*; Ogden, *Swamplife*; Moore, *Suffering for Territory*; Trudeau, "Politics of Belonging"; Walker and Fortmann, "Whose Landscape?"; Tsing, *Mushroom at the End of the World*.
56. I draw on the rich and vast literatures in Marxist feminism and ecofeminism. The works I have been inspired by in particular include Federici, *Caliban and the Witch*; Mies, *Patriarchy and Accumulation*; Shiva, *Staying Alive*; Merchant, *Death of Nature*; Plumwood, *Feminism and the Mastery of Nature*; Stoler, *Race and the Education of Desire*; McClintock, *Imperial Leather*.
57. Butler, *Gender Trouble*, 34.
58. On hegemonic masculinity and femininity as well as multiple masculinities and femininities see Connell, *Men and the Boys*; Connell and Messerschmidt, "Hegemonic Masculinity"; Schippers, "Recovering the Feminine Other."
59. McClintock, *Imperial Leather*.
60. Hancock, *Solidarity Politics for Millennials*.
61. Lugones, "Towards a Decolonial Feminism." See also Mama, "Sheroes and Villains"; Connell, "Sociology of Gender"; Kitunga and Mbilinyi, "Rooting Transformative Feminist Struggles"; Mollett and Faria, "Messing with Gender"; Radcliffe, *Dilemmas of Difference*.
62. Bryceson, "Gender Relations in Rural Tanzania"; Sheriff, *Slaves, Spices and Ivory in Zanzibar*; Cooper, *Plantation Slavery*.
63. Brown, "Politics of Business," 230; Tanzania National Archives (TNA), Tanganyika Territory, District Officer's Reports Bagamoyo District, 1926.
64. Montgomery, "Colonial Legacy of Gender Inequality."
65. Mbilinyi, "'City' and 'Countryside' in Colonial Tanganyika"; Bryceson, "Gender Relations in Rural Tanzania"; Hodgson, *Gender, Justice, and the Problem of Culture*.
66. Freyhold, *Ujamaa Villages in Tanzania*.
67. Mbilinyi, "'City' and 'Countryside' in Colonial Tanganyika," 96.
68. Dancer, *Women, Land and Justice*; Hodgson, *Gender, Justice, and the Problem of Culture*.
69. Coquery-Vidrovich, *African Women*; Connell, "Sociology of Gender."
70. Nyerere, "Ujamaa."
71. Lal, "Maoism in Tanzania," 97.
72. On the postcolonial nationalist rhetoric and significance of the phrase *kujenga taifa* see Brennan, *Taifa*; Scotton, "Some Swahili Political Words."
73. Brennan, "Blood Enemies."
74. CCM was established in 1977 through the merger of the Tanganyika African National Union (TANU) of mainland Tanzania and the Afro-Shirazi Party of Zanzibar. On nationalism and "imagined communities" see Anderson, *Imagined Communities*.
75. On Marx's description of the peasantry see Marx, "Eighteenth Brumaire."
76. Mbilinyi, "Analysing the History of Agrarian Struggles"; Lal, *African Socialism*; Havnevik, *Tanzania: The Limits to Development*; Ibbott, *Ujamaa*.
77. Lal, *African Socialism*. On the legacies of villagization on the relatively gender-equitable land distribution in Kilombero District see Sulle and Dancer, "Gender, Politics, and Sugarcane."

78. Hodgson, *Once Intrepid Warriors*; Aminzade, *Race, Nation, and Citizenship*; Brennan, *Taifa*.

79. Phillips, *Ethnography of Hunger*, 15.

80. Phillips, 16.

81. Chambi Chachage's notion of "statist neoliberalism" ("Land Acquisition and Accumulation," 9) is useful here for describing the current moment in Tanzania, where the state, despite having undertaken sweeping neoliberal reforms, has not formally renounced a core tenet of socialism: the national ownership of land. On the contradiction between accumulation and legitimation undergirding postcolonial nation building see Aminzade, "Dialectic of Nation Building."

82. I elaborate on the proposed outgrower scheme in chapter 1. On value-chain agriculture as a state-sponsored capitalist spatiotemporal fix see McMichael, "Value-Chain Agriculture."

83. Female farmer, interview with the author, Matipwili, Bagamoyo District, March 19, 2016.

84. Male farmer, interview with the author, Bozi, Bagamoyo District, July 27, 2016.

85. Taussig, *Devil and Commodity Fetishism*, 103. See also Ann Stoler's critique of Taussig's treatment of liminality: Stoler, *Capitalism and Confrontation*, xxn18.

86. Gramsci, *Selections from the Prison Notebooks*; Polanyi, *Great Transformation*; Foucault, *History of Sexuality*. Yet, resistance should not be mistaken for direct opposition of power. Resistance and power are intimately interconnected, and the tension between them always leaves room for ambiguity and ambivalence, as numerous scholars have highlighted. See Abu-Lughod, "Romance of Resistance"; Mitchell, "Everyday Metaphors of Power"; Ortner, "Resistance and the Problem of Ethnographic Refusal"; Roseberry, "Hegemony and the Language of Contention."

87. Scott, *Moral Economy of the Peasant*; Scott, *Weapons of the Weak*; Scott, *Domination and the Arts of Resistance*. I also draw on the rich agrarian studies and political ecology literature that further expand and complicate Scott's work. See, for example, Watts, *Silent Violence*; Hart, "Engendering Everyday Resistance"; Peluso, *Rich Forests, Poor People*; Stoler, *Capitalism and Confrontation*; Neumann, *Imposing Wilderness*; Schroeder, *Shady Practices*; Carney, "Gender Conflict in Gambian Wetlands"; Moore, *Suffering for Territory*; Wolford, *This Land Is Ours Now*.

88. On "think description" see Geertz, *Interpretation of Cultures*.

89. Nor is it analytically useful to generalize all peasants as "risk-averse," "profit-maximizing," or "uncaptured" by the state and the market, as some of the most influential writers on agrarian politics have initially assumed. See Scott, *Moral Economy of the Peasant*; Popkin, *Rational Peasant*; Hyden, *Beyond Ujamaa*.

90. On varying perspectives on the usefulness and dangers of strategic essentialism see, for example, Spivak, "Subaltern Studies"; Wolford, *This Land Is Ours Now*; Collins and Bilge, *Intersectionality*. I differentiate between social positions and social positionings here to highlight the ways different people articulate and act upon their embodied positions in society. As Dorothy Hodgson writes, the concept of positionings "encompasses and signals the interlocking struggles over representation, recognition, resources, and rights that are central to any form of political action" (*Being Maasai, Becoming Indigenous*, 8). I suggest that obviating social difference in the study of rural resistance limits our inquiry into how hegemony and the state operate at multiple scales to set the limits on what is politically possible, what kinds of behaviors are considered socially acceptable within prevailing cultural norms, and which actors are considered "suitable" and "unsuitable comrades" in agrarian struggles.

91. Mahmood, "Feminist Theory, Embodiment, and the Docile Agent," 9.

92. Ortner, drawing on Geertz, defines the "ethnographic stance" as an approach to "producing understanding through richness, texture, and detail, rather than parsimony, refinement, and (in the sense used by mathematicians) elegance" ("Resistance and the Problem of Ethnographic Refusal," 174). Building on this work, I argue for a critical ethnographic stance that is at once committed to thick descriptions and social justice.

93. I first worked in Zanzibar as a community development volunteer in 2009 through the Seoul-based Korea International Volunteer Organization and the Dar es Salaam–based youth development organization UVIKIUTA. Between 2010 and 2012 I worked in London as a research and communication officer at the Johannesburg-based ActionAid International.

94. Handwritten signs advertising farmland for sale ("nauza ardhi / shamba lipo") are frequent sightings along the sixty-kilometer drive on the relatively recently completed tarmac road from Dar es Salaam to Bagamoyo.

95. The translation of *mhindi* and the categories "Indian" and "Asian" require careful consideration. In East Africa and in Britain, "Asian" typically refers to people of South Asian descent or those originating from what are now India and Pakistan. After the partitioning of India and the creation of Pakistan in 1947, "Asian" became a more popular term of reference for these groups than "Indian." These distinctions, however, simplify the differences and contestations around class, caste, gender, religion, and linguistic boundaries within these groups. See Brennan, *Taifa*; Aminzade, *Race, Nation, and Citizenship*; Nagar, "South Asian Diaspora."

96. I use "discursive whiteness" to refer to a shifting assemblage of visual and representational signifiers of class, mobility, and appearance that contribute to the social construction of whiteness as a cultural category. Discursive whiteness in Tanzania and other postcolonial settings is not necessarily tied to white bodies or European ancestry but shaped by the legacies of European colonialism that permeate the contemporary cultural and ideological space as well as the ubiquitous presence of Euro-Americans in international development. See also Faria and Mollett, "Critical Feminist Reflexivity"; Shome, "Whiteness and the Politics of Location."

97. I lived a short distance away from the land concession, lest living within its borders would pose undue risks to my research participants and to myself, though I had obtained necessary permissions and research clearance at the district and national level. I traveled daily back and forth between different settlements and villages, except for when I was in Bagamoyo town or Dar es Salaam, conducting interviews, attending policy events, or doing archival research. Physical access to some villages and subvillages was not without challenges. For instance, reaching Matipwili, a village that straddles the northern boundary of the project site, required passing through a gate put up by Saadani National Park. To pass through the gate, I had to either pay the tourist entrance fee every time or obtain a research permit from the Tanzania National Parks Authority, a parastatal agency under the Ministry of Natural Resources and Tourism. I eventually obtained the permit, but I would later learn that Matipwili villagers had rejected the building of the gate and that there were unresolved land disputes not only between the state, the foreign investor, and the villagers, but also between different agencies within the state, the details of which I discuss in chapter 2.

98. I initially hoped to interview each adult family member in every household, but it proved to be difficult in some instances where husbands or fathers played a strong gatekeeping role.

99. I chose to integrate photovoice into my research for several reasons, including my prior experience with the method, the limits of sedentary interviews, and the frequency with which people wished to show me various things that came up during our conversations. On photovoice and photo elicitation see, for example, Wang and

Burris, "Photovoice"; Harper, "Talking about Pictures"; Chung, Young, and Bezner Kerr, "Rethinking the Value of Unpaid Care Work."

100. Prior to selecting thirty images, the photovoice participants and I discussed all of their images, which I projected onto a tablet. This required on average four visits with each participant, with each visit lasting about sixty to ninety minutes.

1. THE MAKING OF A SWEET DEAL

1. Agro EcoEnergy Tanzania, "White Paper," 3.

2. On theoretical and methodological considerations on studying "the state" see, for example, Abrams, "Notes on the Difficulty"; Mitchell, "Limits of the State"; Wolford et al., "Governing Global Land Deals."

3. The three energy companies include Övik Energi, Umeå Energi, and Skellefteå Kraft. See Sekab, "Our History."

4. Agro EcoEnergy Tanzania, "White Paper," 3.

5. An unpublished internal Sida document, titled "Shareholder Structure"; Foresight Group, "Göran Carstedt."

6. EU, Directive 2003/30/EC; EU, Directive 2009/28/EC; Kingdom of Sweden, "Integrated Climate and Energy Policy."

7. Sekab, "Ethanol as a Raw Material."

8. On crisis tendencies of capitalism and various "fixes" see Harvey, *Limits to Capital*; McCarthy, "Socioecological Fix."

9. Carstedt, "Sustainable Ethanol," 55.

10. Carstedt, 55.

11. Carstedt, 57.

12. Sida representative, interview with the author, Dar es Salaam, October 26, 2015.

13. Personal communication with a Swedish researcher, April 4, 2016.

14. Hittapunktse AB, "Carstedts Bil AB." An internal Sida document cited in note 5 indicates that Carstedts Bil AB is owned by Carstedts Service AB, which in turn is owned by Ecosystem i Scandinavien AB (Ecosystems), a majority shareholder of EcoDevelopment.

15. Kamanga, "Agrofuel Industry." The BioAlcohol Fuel Foundation which Per Carstedt also chairs, was established in 1983 to facilitate technological advancements in the production and use of biofuels. No public information is available on the Community Finance Company (CFC), but according to interviews with key informants, one of its owners, Amani Sinare, and members of his family hold significant political clout in Tanzania. Sinare's sister, Mwanaidi Sinare Maajar, for instance, served as the Tanzanian ambassador to the US between 2010 and 2013 and as the high commissioner to the UK between 2006 and 2010, both under the Kikwete administration. A 2010 report published by HakiArdhi, a land rights research institute based in Dar es Salaam, described CFC as "merely a broker" whose role was to facilitate the initial meetings between Swedish entities and high-profile Tanzanian authorities. See Chachage, "Land Acquisition."

16. GTZ, *Liquid Biofuels*, 122.

17. GTZ, 6.

18. Kojima and Johnson, "Potential for Biofuels," 11.

19. Members of the task force included the following ministries: Energy and Minerals; Planning, Economy, and Empowerment; Agriculture, Food, and Cooperatives; Finance; Labour, Employment, and Youth Development; Water and Irrigation; Lands, Housing, and Human Settlements Development; plus the Vice President's Office (Division of Environment); the Attorney General's Chambers; the Tanzania Investment Centre; the Tanzania Petroleum Development Corporation; the CFC; and the Tanzania Sugar Producers' Association. See Kamanga, "Agrofuel Industry."

20. Sekab BT, "Environmental and Social Impact Statement," 46.
21. Sekab BT was owned 98.5 percent by Sekab and 1.5 percent by CFC.
22. Havnevik and Haaland, "Biofuel, Land and Environmental Issues."
23. ActionAid Tanzania, "Implications of Biofuels"; ActionAid Sweden, "Sekab"; WWF Tanzania, "Biofuel Industry"; Roberntz, Edman, and Carlson, "Rufiji Landscape."
24. Havnevik and Haaland, "Biofuel, Land and Environmental Issues."
25. Roberntz, Edman, and Carlson, "Rufiji Landscape." For similar critiques of biofuels see Holt-Giménez, "Biofuel Myth"; Gallagher, "Gallagher Review."
26. PMO official, interview with the author, Dar es Salaam, October 29, 2015.
27. PMO official interview.
28. URT, Land Act, 1999.
29. URT, Land Act, 1999, Part 1, Section 2.
30. On rendering land investible see Li, "What Is Land?"
31. URT, Tanzania Investment Act, 1997.
32. TIC official, interview with the author, Dar es Salaam, November 16, 2015; URT, Village Land Act, 1999, Part 3, Section 4(6).
33. TIC official interview; URT, Village Land Act, 1999, Part 3, Section 4(6).
34. URT, Village Land Act, 1999, Part 3, Section 4.
35. Shivji, "Lawyers in Neoliberalism."
36. TIC official interview. In fact, one study found that in 2008 alone, the TIC received 270 applications for land by investors but already had a backlog of over 4,000 applications. See OECD, "Investment Policy Reviews."
37. Assistant commissioner for lands (legal services unit), Ministry of Lands, interview with the author, Dar es Salaam, August 1, 2016.
38. Sekab BT, "Environmental and Social Impact Statement," 50.
39. Per Carstedt, interview with the author, Dar es Salaam, August 22, 2014.
40. Carstedt interview, August 22, 2014.
41. Sekab BT employee, interview with the author, Bagamoyo town, March 18, 2016.
42. Carstedt interview, August 22, 2014.
43. In the midst of the global land rush, Sekab BT was frequently featured in media reports, NGO publications, policy documents, and scholarly works as a case study of land grabbing in Africa. See, for example, ActionAid Tanzania, "Implications of Biofuels"; ActionAid Sweden, "Sekab"; WWF Tanzania, "Biofuel Industry"; Roberntz, Edman, and Carlson, "Rufiji Landscape"; Oakland Institute, "Understanding Land Investment Deals in Africa"; Sulle and Nelson, "Biofuels, Land Access"; Cotula, "Land Grab or Development Opportunity?"
44. *Development Today*, "Filed Complaint"; *Development Today*, "Local Judge."
45. *Development Today*, "Sekab 'Substantially Altered.'"
46. For a thorough review on the ESIA controversy see Havnevik, Haaland, and Abdallah, "Biofuel, Land, and Environmental Issues."
47. *Development Today*, "Sekab 'Substantially Altered.'"
48. Sida representative interview.
49. Havnevik and Haaland, "Biofuel, Land and Environmental Issues."
50. Humphrey and Prizzon, "Guarantees for Development," 6; Halvorson-Quevedo and Mirabile, "Guarantees for Development."
51. Sekab BT, "Application," 1.
52. Sekab BT, "Application," 2.
53. Havnevik and Haaland, "Biofuel, Land and Environmental Issues."
54. Sida, "Letter Re: Sekab," 2009, 2.
55. Sida representative interview.
56. *Development Today*, "Internal Report."

57. On the FAQ section of Sekab's corporate web page in 2009, which has since been taken down, the company explained the Mauritian establishment as follows: "Previously, in 2010 and at the request of some large institutional investors, EcoEnergy established an African holding company in Mauritius, with the purpose of facilitating entry for some strategic investors on the EcoEnergy African holding company level."

58. According to the Swedish Companies Registration Office, as of 2022, Per Carstedt owned more than 25 percent but less than 50 percent shares in both EcoDevelopment in Europe AB and EcoEnergy Africa AB.

59. I draw the shareholding information from an unpublished Sida report, "Current Owners of Bagamoyo EcoEnergy," dated October 2014.

60. Sekab BT employee interview.

61. Existing sugar mills include the Kilombero Sugar Company, Mtibwa Sugar Estates, the Tanganyika Planting Company, and Kagera Sugar. See Sulle, "Social Differentiation."

62. SAGCOT, "SAGCOT Investment Blueprint"; Rabobank, "Tanzania Sugar."

63. Sulle, "Social Differentiation."

64. Ministry of Agriculture official, interview with the author, Dar es Salaam, October 16, 2015.

65. PMO official interview.

66. Per Carstedt, interview with the author, Dar es Salaam, October 30, 2015. While EcoEnergy would focus primarily on producing processed sugar for the Tanzanian market, it still planned to generate electricity and ethanol from bagasse and molasses, byproducts of sugar refining. Electricity would be supplied to the Tanzanian national grid, and ethanol would be purchased by Sekab in Sweden. As part of an off-take contract Sekab signed with EcoDevelopment during the sale of Sekab BT, the Swedish company agreed to purchase all ethanol produced by EcoEnergy for the first ten years of its operation. See Havnevik and Haaland, "Biofuel, Land and Environmental Issues"; Kahoho, "Ethanol."

67. Borras et al., "Rise of Flex Crops"; Hunsberger and Alonso-Fradejas, "Discursive Flexibility"; McKay et al., "Political Economy of Sugarcane Flexing."

68. Sekab BT, "Environmental and Social Impact Statement," 229.

69. Agro EcoEnergy Tanzania, "White Paper," 4.

70. Agro EcoEnergy Tanzania, "White Paper," 20.

71. Agro EcoEnergy Tanzania, "Bagamoyo EcoEnergy."

72. Agro EcoEnergy Tanzania, "White Paper," 4–5.

73. Outgrower specialist / EcoEnergy contractor, interview with the author, Bagamoyo town, August 19, 2013. Here I am reminded of Anna Tsing's assertion that "no firm has to personally invent patriarchy, colonialism, war, racism, or imprisonment, yet each of these is privileged in supply chain labor mobilization." See Tsing, "Supply Chains," 151.

74. IMF, "United Republic of Tanzania."

75. URT, Budget Digest 2004/5.

76. Cooksey, "Politics, Patronage and Projects."

77. The National Strategy for Growth and the Reduction of Poverty is more commonly known by its Swahili acronym, MKUKUTA (*Mkakati wa Kukuza Uchumi na Kupunguza Umaskini*).

78. Velde and Massa, "Donor Responses."

79. World Bank, *World Development Report 2008*.

80. URT, Kilimo Kwanza Resolution, 1–2.

81. URT, Kilimo Kwanza Resolution, 1–2.

82. WEF, "Realizing a New Vision," 23.

83. SAGCOT, "SAGCOT Investment Blueprint," 4.

84. URT, Tanzania Development Vision 2025.

85. WEF, "Grow Africa," 4.
86. New Alliance, "New Alliance," 2.
87. New Alliance, 1.
88. Cooksey, "What Difference."
89. Jenkins, "Mobilizing."
90. Jenkins, "Mobilizing."
91. Cooksey, "What Difference"; Sulle, "Bureaucrats, Investors and Smallholders." As I return to in the conclusion, these initiatives that gave EcoEnergy its political raison d'être would eventually be dissolved by Kikwete's successor, John Pombe Magufuli. See *Citizen*, "Unanswered Questions."
92. IFAD, "President's Report."
93. These two reports were approved and published on the AfDB website in late August 2012.
94. AfDB representative, interview with the author, Dar es Salaam, May 24, 2016.
95. URT, Hotuba ya Wizara ya Ardhi. Tibaijuka is a prominent CCM politician who had studied agricultural economics in Sweden and served as the under secretary-general of the UN and the executive director of UN-Habitat.
96. URT, Hotuba ya Wizara ya Ardhi.
97. Ashura Yussuf, "Speech on New Land Policy."
98. Letter from Anna Tibaijuka to Per Carstedt, dated February 23, 2012. Ref: CBC 171/234/01/11 on Exchange of Land for Equity in Expansive Land Based Investments, 1.
99. PMO official interview; deputy permanent secretary, Ministry of Lands, interview with the author, Dar es Salaam, August 1, 2016.
100. USAID / World Bank consultant, interview with the author, Dar es Salaam, November 11, 2015.
101. Assistant commissioner for lands interview.
102. Deputy permanent secretary interview.
103. PMO official interview.
104. Vhugen et al., "Approaches," 3, 7, 28.
105. URT, Land Act, 1999, Part 2, Section 1(g); URT, Village Land Act, 1999, Part 3, Section 8(a).
106. Assistant commissioner for lands interview.
107. According to a district land surveyor, the discrepancy in the size of the land area promised to EcoEnergy since 2008 is due to factors such as road expansions and river erosions.
108. Agro EcoEnergy Tanzania, "White Paper."
109. Sida representative interview.
110. Agro EcoEnergy Tanzania, "White Paper"; Sida representative interview.
111. Sida, "Decision on Contribution to EcoEnergy Tanzania."
112. Agro EcoEnergy Tanzania, "White Paper"; AfDB representative interview.
113. AfDB representative interview; Sida representative interview.
114. Agro EcoEnergy Tanzania, "White Paper."
115. Sida representative interview; Sida, "Sida Avslutar Stödet till EcoEnergy"; Songa, "Key Financiers Drop."
116. Sida representative interview.
117. ActionAid International, *Take Action*.
118. Agro EcoEnergy Tanzania, "Press Release"; Agro EcoEnergy Tanzania, "Response to ActionAid."
119. ActionAid Tanzania staff, interview with the author, Dar es Salaam, April 10, 2015. Beyond ActionAid, EcoEnergy had become a cause célèbre for other civil society groups

like Biovision, who, as WWF did several years prior, politicized the environmental consequences of plantation agriculture. See Biovision, "Can Agri-business."

120. IFAD, "President's Report," 11.
121. Assistant commissioner for lands interview.
122. Kabendera and Anderson, "Tanzania Energy Scandal."
123. Policy Forum, "Tanzania Governance Review 2014"; Policy Forum, "Tanzania Governance Review 2015–16."
124. PMO official interview.
125. AfDB representative interview.
126. AfDB representative interview.
127. *Development Today*, "Internal Report."
128. Carstedt interview, October 30, 2015.
129. Li, "What Is Land?"; Engström and Hajdu, "Conjuring a 'Win-World.'"
130. Polanyi, *Great Transformation*.

2. THE MAKING OF A BITTER LANDSCAPE

1. For nineteenth-century economic history of Bagamoyo and the Swahili coast see, for example, Fabian, *Making Identity*; Glassman, *Feasts and Riot*; Sheriff, *Slaves, Spices and Ivory*; Cooper, *Plantation Slavery*.
2. Fabian, "East Africa's Gorée."
3. Lindström, *Muted Memories*.
4. Though enslaved people were traded in Bagamoyo, the epicenter of the nineteenth-century slave trade was Kilwa, farther south along the coastline. See Fabian, "East Africa's Gorée"; Lindström, *Muted Memories*; Rockel, *Carriers of Culture*.
5. Brown, "Politics of Business"; Fabian, *Making Identity*.
6. Brown, "Pre-colonial History of Bagamoyo," 203. So lucrative was the business of the Holy Ghost Fathers that a district officer observed in 1926 that "religion must be a secondary consideration"; TNA / Bagamoyo District Annual Report, 1926, 11.
7. For a discussion on the origins of the name "Bagamoyo" see Brown, "Bagamoyo: Inquiry"; Fabian, "East Africa's Gorée."
8. Notable works about Bagamoyo's urban history include Fabian, *Making Identity*; Brown, "Pre-colonial History of Bagamoyo."
9. GLOWS-FIU, "Water Atlas."
10. Stoler, "Imperial Debris."
11. Mitchell, *Landscape and Power*; Neumann, *Imposing Wilderness*; Ogden, *Swamplife*; Moore, *Suffering for Territory*; Trudeau, "Politics of Belonging"; Walker and Fortmann, "Whose Landscape?"
12. The notion of "presence" has been usefully elaborated by several scholars in the South African context. See Dlamini, *Safari Nation*; Makhulu, *Making Freedom*; Ferguson, *Give a Man a Fish*.
13. District land officer, interview with the author, Bagamoyo town, July 21, 2014; Makurunge village chairman, interview with the author, Makurunge, Bagamoyo District, August 15, 2013.
14. Brown, "Bagamoyo: An Historical Introduction"; Jerman, *Between Five Lines*; Fabian, *Making Identity*.
15. Village elders, interview with the author, Matipwili, Bagamoyo District, September 4, 2016. Edward Alpers also cites an interview he conducted in Matipwili in 1972, in which the story of Funditambuu was central to people's oral history. See Alpers, *Ivory and Slaves*, 33 n27.

16. Village elders interview, Matipwili; village elders, interview with the author, Makaani, Bagamoyo District, March 9, 2016.

17. On *ndago* see Stoller and Wax, "Yellow Nutsedge"; Stoller and Sweet, "Biology and Life Cycle"; Defelice, "Yellow Nutsedge"; Wilen, McGiffen, and Elmore, "Nutsedge."

18. Juma, interview with the author, Matipwili, Bagamoyo District, September 4, 2016.

19. Mwajuma, interview with the author, Matipwili, Bagamoyo District, June 11, 2016.

20. Mwajuma interview.

21. There are over 120 ethnic groups in Tanzania, 80 percent of which are known to be patrilineal (where inheritance and descent is determined through the male line), and the remaining groups, mostly in coastal regions, are traditionally matrilineal. Scholars have attributed the shift among matrilineal groups to patrilineal patterns of inheritance to the influence of Islam and the migration of patrilineal ethnic groups to the coast, as well as the expansion of Christianity and the colonial policies that normalized the ideal of the patriarchal nuclear family. See Beidelman, *Matrilineal Peoples*; Dondeyne et al., "Changing Land Tenure Regimes"; Koda, "Changing Land Tenure Systems"; Bryceson, "Gender Relations in Rural Tanzania"; Englert, "From a Gender Perspective."

22. Focus group discussion, Razaba, Bagamoyo District, August 16, 2013.

23. As noted in the glossary, people often used "kitopeni" as both a particular season and a geographic area.

24. Field notes, May 2016.

25. Farmers, interviews with the author, Matipwili, Bagamoyo District, February 20, 2016; September 4, 2016; October 31, 2016; June 23, 2016. See also *BBC News*, "Indian Ocean Dipole."

26. Field notes, July 2016.

27. Mwajuma interview.

28. Elsewhere I argue for a conceptualization of land and water as a coupled resource. See Chung, "Grass Beneath."

29. Village elders, interview with the author, Matipwili, Bagamoyo District, August 5, 2014.

30. Booth and Wickens, "Non-timber Uses"; Hines and Eckman, "Indigenous Multipurpose Trees."

31. On anthropological studies of these matrilineal ethnic groups see Beidelman, *Matrilineal Peoples*; Swantz, *Ritual and Symbol*; Swantz, *Blood, Milk, and Death*; Jerman, *Between Five Lines*; Vuorela, *Women's Question*.

32. Anthropologist Marja-Liisa Swantz described mkole as "the tree of greatest symbolic significance" for Zaramo women, though my research shows that her observation applies to other matrilineal coastal ethnic groups in Bagamoyo. Swantz, *Blood, Milk, and Death*, 140.

33. Swantz, *Blood, Milk, and Death*; Thompson, "Embodied Socialization."

34. Mwanahamisi, interview with the author, Gama/Matipwili, Bagamoyo District, January 26, 2016.

35. Female farmer, interview with the author, Bozi, Bagamoyo District, July 23, 2014.

36. Male farmers, interviews with the author, Matipwili, Bagamoyo District, November 28, 2015; January 13, 2015; field notes, December 2015. *Maulidi* (from Arabic, *mawlid*) are religious celebrations to commemorate the prophet Muhammad's birth, but they may be organized for other special occasions, including weddings, funerals, female initiation ceremonies, and during Ramadan.

37. I use "RAZABA" to denote the ranch and "Razaba" to refer to the subvillage that was established after the ranch's closure.

38. Male elder, interview with the author, Matipwili, Bagamoyo District, March 15, 2016.

39. Calvert, *German East Africa*; Sunseri, "Baumwollfrage"; Iliffe, *Modern History of Tanganyika*.
40. Sunseri, "Baumwollfrage."
41. Sunseri, "Famine and Wild Pigs."
42. Sunseri, "Baumwollfrage."
43. Focus group discussion and participatory mapping with village elders, Matipwili, Bagamoyo District, March 5, 2016.
44. Sunseri, "Baumwollfrage."
45. The name Kisauke appears in the earliest colonial maps, dated back to the early 1900s.
46. Focus group discussion and participatory mapping, Matipwili, March 5, 2016.
47. Sunseri, "Baumwollfrage," 47.
48. Iliffe, *Modern History of Tanganyika*.
49. Field notes, March 2016.
50. Iliffe, *Modern History of Tanganyika*, 145.
51. TNA / Bagamoyo District Annual Report, 1926; Tenga and Mramba, "Manual on Land Law."
52. Tanganyika Territory, Land Ordinance, 1923, Section 3.
53. James and Fimbo, *Customary Land Law*, 7.
54. TNA / Bagamoyo District Annual Report, 1925, 6.
55. Focus group discussion and participatory mapping, Matipwili, March 5, 2016.
56. TNA / Extract from Annual Report Eastern Province by G. F. Webster, Esq., 1926.
57. Male elder, interview with the author, Makaani, Bagamoyo District, August 11, 2014.
58. TNA / Bagamoyo District Annual Report, 1931.
59. TNA / Bagamoyo District Annual Report, 1936.
60. Male elder, interview with the author, Matipwili, Bagamoyo District, August 11, 2014.
61. Male elder, interview with the author, Matipwili, Bagamoyo District, November 28, 2015.
62. Female elder, interview with the author, Gama/Matipwili, Bagamoyo District, March 9, 2016.
63. Shivji, *Not Yet Democracy*.
64. Nyerere, "Ujamaa," 17.
65. Hyden, *Beyond Ujamaa*; Askew, "Sung and Unsung"; Scott, *Seeing Like a State*.
66. Ergas, "Why Did the Ujamaa Village Policy Fail?"
67. McHenry, *Tanzania's Ujamaa Villages*; Lal, *African Socialism*.
68. Male elder, interview with the author, Matipwili, Bagamoyo District, January 13, 2015.
69. Male elder, interview with the author, Matipwili, Bagamoyo District, September 4, 2016.
70. Field notes, September 2016.
71. According to government telegrams dated 1976 and 1977, the size of this land concession was 77,663 acres (31,429 hectares). Survey maps of the ranch in 1985 and 1988 give different measurements at 26,297 and 28,097 hectares respectively. The district land officer attributes these discrepancies to coastal erosion and road expansion. District land officer interview; land rights activist, interview with the author, Bagamoyo town, August 19, 2016.
72. On union politics see Chase, "Zanzibar Treason Trial"; Aminzade, *Race, Nation, and Citizenship*, 201–4.
73. Male elder, interview with the author, Razaba, Bagamoyo District, August 19, 2016.

74. Telegram from Viwanja–Dar es Salaam to Kilimo-Zanzibar, dated February 21, 1977, Kuh: Maombi ya Kupatiwa Ardhi Ekari 77,663 ya Wilayani Bagamoyo. Saving No. ID/14429/135/VB.

75. The original phrase in the letter read "vitu vilivyomo, pamoja na watu na haki zao zisisumbuliwe kwa hali yo yote ile."

76. RAZABA manager, interview with the author, Bagamoyo town, February 11, 2015.

77. Focus group discussion and participatory mapping, Matipwili, March 5, 2016.

78. Tanzania High Court (Land Division) at Dar es Salaam. Land Case No. 275 of 1989 (unpublished).

79. Letter from Jakaya Kikwete, minister for water, energy, and mineral resources, to prime minister, January 3, 1994.

80. Land rights activist interview.

81. Former RAZABA workers, interviews with the author, Razaba, Bagamoyo District, December 21, 2015, and August 19, 2016.

82. Male elder, interview with the author, Razaba, Bagamoyo District, August 19, 2016; Sekab BT, "Environmental and Social Impact Statement," ix.

83. "Bozi" literally means a fool in Swahili (pl., *mabozi*). As an elder resident explained, the settlement was so named because of a dam there that would periodically swell and dry up in ways that confused people.

84. Former RAZABA workers interviews.

85. RAZABA manager interview.

86. RAZABA manager interview.

87. Male farmer, interview with the author, Razaba, Bagamoyo District, August 30, 2016.

88. On the dispossession of the Barabaig see Tenga and Kakoti, "Barabaig Land Case"; Lane, *Pastures Lost*.

89. Shivji, *Not Yet Democracy*; Tsikata, "Securing Women's Interests"; McAuslan, *Land Law Reform*.

90. In practice, however, customary rights are often assumed to be inferior to statutory rights, a tendency inherited from the colonial period. See James and Fimbo, *Customary Land Law*; Tenga and Mramba, "Manual on Land Law."

91. URT, Village Land Act, 1999, Part 4, Section 57(1).

92. TANAPA was established by the Tanganyika National Parks Ordinance Cap. 412 of 1959 and is governed by the National Parks Act Chapter 282 of 2002 and the Wildlife Conservation Act No. 5 of 2009.

93. URT, Wildlife Conservation (Game Reserves) Order, 1974 (G.Ns. Nos. 265 and 275 of 1974); Baldus, Roettcher, and Broska, "Saadani."

94. Neumann, *Imposing Wilderness*. According to Orozco-Quintero and King ("Cartography of Dispossession"), negotiations between Saadani village and the Wildlife Division began in 1965.

95. Baldus, Roettcher, and Broska, "Saadani."

96. Baldus, Roettcher, and Broska, 9.

97. Neumann, *Imposing Wilderness*.

98. Nash, *Wilderness and the American Mind*, 342.

99. Focus group discussion and participatory mapping, Matipwili, March 5, 2016.

100. IRA, "Environmental Evaluation."

101. Male elder interview, Matipwili, Bagamoyo District, July 30, 2016. GN No. 265 and 275 are clear in their boundary descriptions that the game reserve leaves "Kisauke Sisal Factory" and "Maguke Village [sic]" to the south.

102. Focus group discussion and participatory mapping, Matipwili, March 5, 2016.

103. Baldus, Roettcher, and Broska, "Saadani."
104. TANAPA, "Saadani National Park."
105. Booth, "Options and the Future Status," 1.
106. Focus group discussion and participatory mapping, Matipwili, March 5, 2016.
107. Matipwili village archive, letter from the Department of Natural Resources, Office of the District Executive Director, to the ward executive director, dated July 2, 1999, regarding the recovery of the Saadani Game Reserve boundary (Taarifa ya Ufuatiliaji Mpaka wa Pori la Akiba la Wanyama Pori Saadani, Kumb. Na. CD. 80/DC/BG/13 . . . /SM).
108. Matipwili village archive, letter from the Department of Natural Resources, Office of the District Executive Director, to the ward executive director, dated July 2, 1999, regarding the recovery of the Saadani Game Reserve boundary, 17.
109. Village elder, interview with the author, Matipwili, Bagamoyo District, June 10, 2016.
110. Female elder, interview with the author, Matipwili, Bagamoyo District, August 15, 2014.
111. Former village leader, interview with the author, Matipwili, Bagamoyo District, September 3, 2016.
112. Female elder interview, Matipwili, August 15, 2014.
113. Matipwili village archive, "Various steps followed in establishing the Saadani National Park," n.d.
114. The fact that compensation payments were done at night is also documented in an unpublished letter from the Bagamoyo district commissioner, Hawa S. Ngulume, to the TANAPA director general, dated March 12, 2005, re: Conflict between TANAPA and citizens in the villages of Saadani, Matipwili, and Mkange, in Mkange Ward, Miono Division [Mgogoro kati ya TANAPA na Wananchi Tarafa ya Miono Kata ya Mkange, Vijiji Vya Saadani, Matipwili-Java na Mkange, Kumb. Na. BG/G.10/2/9].
115. Matipwili village archive, "Various steps followed in establishing the Saadani National Park," n.d.
116. RAZABA manager interview.
117. URT, Declaration of Saadani National Park, Government Notice No. 281.
118. Male elder, interview with the author, Matipwili, Bagamoyo District, July 26, 2016.
119. Field notes, September 2014.
120. Female elder, interview with the author, Matipwili, Bagamoyo District, July 27, 2014.
121. Per Carstedt, interview with the author, Dar es Salaam, October 30, 2015.

3. ON BEING COUNTED

1. Operational policies on involuntary resettlement emerged in the late 1970s in response to major World Bank–funded dam and infrastructure-related development projects that resulted in mass displacement and impoverishment of local populations. The earliest policy statement regarding involuntary resettlement, called "Social Issues Associated with Involuntary Resettlement" (Operational Manual Statement, OMS 2.33), dates back to 1980 and was drafted by a sociologist, Michael Cernea. The policy was subsequently revised and updated in the 1990s and 2000s and was adopted by other multilateral development banks and bilateral aid agencies.
2. IFC, "Performance Standard 5," 1; AfDB, "Integrated Safeguards System," 31. On critique of involuntary resettlement and the IFI's definition of displacement (as physical and economic displacement) see Chung, "Engendering the New Enclosures." I would also refer readers to the literature on development-induced displacement and resettlement,

including but not limited to Colson, *Social Consequences of Resettlement*; Fahim, *Dams, People, and Development*; McDowell, *Understanding Impoverishment*; Indra, *Engendering Forced Migration*; Cernea and McDowell, *Risks and Reconstruction*.

3. See, for example, Askew and Odgaard, "Deeds and Misdeeds"; Lastarria-Cornhiel, "Impact of Privatization"; Deere and León, "Gender Asset Gap"; Radcliffe, "Gendered Frontiers."

4. Agro EcoEnergy Tanzania, "Draft Resettlement Action Plan," 141.

5. Female farmer, interview with the author, Matipwili, Bagamoyo District, August 21, 2013.

6. Male farmer, interview with the author, Matipwili, Bagamoyo District, July 25, 2014.

7. Senior government official in the Ministry of Lands, Housing, and Human Settlements Development, interview with the author, Dar es Salaam, September 28, 2015.

8. URT, Land Acquisition Act, 1967; URT, Constitution of the United Republic of Tanzania, 1977; URT, Land Act, 1999.

9. URT, Land Act, 1999; URT, Land (Assessment of the Value of Land for Compensation) Regulations, 2001.

10. See, for example, Kironde, "Improving Land Sector Governance"; Kombe, "Land Acquisitions for Public Use"; Pedersen and Kweka, "Political Economy of Petroleum Investments."

11. URT, Village Land Act, 1999.

12. Chief valuer, Ministry of Lands, Housing and Human Settlements Development, interview with the author, Dar es Salaam, August 18, 2016.

13. *Citizen*, "Dar Land Sharks."

14. Kombe, "Land Acquisitions for Public Use."

15. Agro EcoEnergy Tanzania, "Draft Resettlement Action Plan," 147.

16. The Swahili words people used were *kutathminiwa* and *hakutathminiwa*. They are literally translated, respectively, as "evaluated" and "not evaluated," but for a smoother translation, I am using "counted" and "not counted."

17. The estimated compensation value, which was never revealed to local residents, ranged from TZS 0 to about 20 million (USD 10,000), with a median of approximately TZS 260,000 (USD 130).

18. Land Form 69a is a version of Land Form 69; the form can be modified to address particular circumstances of each case. See URT, Land (Compensation Claims) Regulations, 2001.

19. Chung, "Engendering the New Enclosures."

20. Chief valuer interview.

21. Chief valuer interview.

22. Senior government official in the Ministry of Lands, Housing, and Human Settlements Development, interview, September 28, 2015.

23. Field notes, July 2014.

24. Foreign development consultant, interview with the author, Bagamoyo town, August 9, 2013.

25. See, for example, Mehta, "Double Bind"; Colson, "Gendering Those Uprooted"; Cornish and Ramsay, "Gender and Livelihoods in Myanmar."

26. Kironde, "Regulatory Framework."

27. Young female farmer, interview with the author, No. 4, Bagamoyo District, July 20, 2016.

28. Middle-aged female farmer, interview with the author, Makaani, Bagamoyo District, August 11, 2014.

29. Friedmann, "Patriarchy and Property," 192. On feminist critiques of the neoclassical model of the household in an agrarian context see, for example, Agarwal, "'Bargaining' and Gender Relations"; Kabeer, *Reversed Realities*; Whitehead, "I'm Hungry Mum"; Guyer and Peters, "Conceptualizing the Household"; Carney and Watts, "Manufacturing Dissent"; Schroeder, *Shady Practices*.

30. Sheriff, *Slaves, Spices and Ivory*; Rockel, *Carriers of Culture*; Fabian, *Making Identity*.

31. Amina, interview with the author, Bozi, Bagamoyo District, June 27, 2016.

32. Rukia, interview with the author, Makaani, Bagamoyo District, August 11, 2014. Encounters like this bring into sharp relief the uneven power dynamics between the researcher and "the researched" and the racialized assumptions of authority that prevail in the field. These moments also reinforce the importance of critical reflexivity as integral to the practice of ethnography.

33. She is also the author of the photo-narrative featured in the introduction (figure 0.5).

34. Nuru, interview with the author, Matipwili, Bagamoyo District, July 25, 2014.

35. Nuru interview, July 25, 2014.

36. Nuru, interview with the author, Matipwili, Bagamoyo District, September 28, 2016. The English equivalent of the proverb might be "A bird in hand is worth two in the bush."

37. De Soto, *Mystery of Capital*; Deininger, "Land Policies for Growth"; Byamugisha, "Securing Africa's Land."

38. World Bank, "Tanzania: New World Bank Financing." For critiques of donor-funded land formalization programs in Tanzania see Askew and Odgaard, "Deeds and Misdeeds"; Maganga et al., "Dispossession through Formalization"; Stein et al., "Formal Divide"; Bluwstein et al., "Between Dependence and Deprivation." For feminist critiques of land formalization in Africa, Asia, and Latin America see, for example, Whitehead and Tsikata, "Policy Discourses"; Peters, "Our Daughters"; Yngstrom, "Women, Wives, and Land Rights"; Meinzen-Dick and Mwangi, "Cutting the Web of Interests"; Beban, *Unwritten Rule*; Deere and León, "Who Owns the Land?"; Lastarria-Cornhiel, "Impact of Privatization."

39. Chief valuer interview.

40. Per Carstedt, interview with the author, Dar es Salaam, October 30, 2015.

41. Carstedt interview.

4. GOVERNING LIMINALITY

1. Field notes, July 2014.
2. Mosi, interview with the author, Bozi, Bagamoyo District, July 28, 2014.
3. Agro EcoEnergy Tanzania, "Draft Resettlement Action Plan," 160.
4. Agro EcoEnergy Tanzania, "Draft Resettlement Action Plan," 24.
5. AfDB, "Involuntary Resettlement Policy," 10, emphasis in original.
6. Foucault, *Society Must Be Defended*, 248.
7. Development consultant (Anna), interview with the author, Bagamoyo District, November 16, 2015.
8. Development consultant's intern, interview with the author, Bagamoyo District, July 26, 2014.
9. Development consultant's intern interview.
10. Agro EcoEnergy Tanzania, "Draft Resettlement Action Plan," 24.
11. Development consultant (Anna), interview with the author, Bagamoyo town, August 9, 2013.
12. Carpentry trainee #1, interview with the author, Bagamoyo District, August 13, 2013.

13. Driving trainee, interview with the author, Bagamoyo District, August 13, 2013.
14. Entrepreneurship trainee #1, interview with the author, Bagamoyo District, August 1, 2014.
15. Carpentry trainee #2, interview with the author, Bagamoyo District, August 13, 2013.
16. Entrepreneurship trainee #2, interview with the author, Bagamoyo District, July 28, 2014.
17. Catering trainee, interview with the author, Bagamoyo District, August 16, 2013.
18. Male farmer, interview with the author, Bozi, Bagamoyo District, August 3, 2016.
19. English literacy trainee #1, interview with the author, Bagamoyo District, September 7, 2016.
20. English literacy trainee #2, interview with the author, Bagamoyo District, August 22, 2013.
21. Zainab and Yusuf, interview with the author, Bozi, Bagamoyo District, August 14, 2013.
22. Zainab, interview with the author, Mkwajuni, Bagamoyo District, July 23, 2014.
23. Zainab interview.
24. Field notes, September 2015.
25. Halima, interview with the author, Bagamoyo District, July 2, 2016.
26. Anna estimated that EcoEnergy had owed her USD 40,000; by early 2016 she was preparing to take the company to court.
27. Brennan, "Youth, the TANU Youth League," 235.
28. Aminzade, *Race, Nation, and Citizenship*.
29. Lal, *African Socialism*.
30. Shaidi, "Crime, Justice and Politics"; Lal, *African Socialism*.
31. URT, People's Militia (Powers of Arrest) Act, 1975.
32. Shivji, *State Coercion and Freedom*, 70 and 67.
33. On the prerogative powers or the "masculinism of the state" see Brown, *States of Injury*, 167.
34. Mbembe, "Necropolitics," 39.
35. Mgambo recruit, interview with the author, Bagamoyo District, December 1, 2015.
36. Cornwall, "To Be a Man"; Cross, "Community Policing."
37. Mgambo recruit interview.
38. Saidi, interview with the author, Bagamoyo District, March 18, 2016.
39. On the prerogative powers of district and regional governments see URT, Regional Administration Act, 1997.
40. Young, "Logic of Masculinist Protection."
41. Field notes, various dates, 2015–2016.
42. Meena, interview with the author, Bagamoyo District, March 18, 2016.
43. Field notes, June 2016.
44. Selemani, interview with the author, Makaani, Bagamoyo District, December 10, 2016.
45. Zuberi and Beatrice, interview with the author, Bagamoyo District, September 7, 2016. According to Anna, Ali used to be her driver and was later hired by EcoEnergy as a "field manager."
46. Saidi, interview with the author, Bagamoyo town, December 1, 2015.
47. When I met Per Carstedt in late 2016, he neither acknowledged nor denied the mgambo violence, and noted that EcoEnergy was "a responsible company that respects human rights."
48. Scott, "Evidence of Experience."
49. Daudi, interview with the author, Bagamoyo District, March 18, 2016.

50. Shabani, interview with the author, Bagamoyo District, March 18, 2016.
51. Daudi interview.
52. Shabani interview.
53. Christina, interview with the author, Bagamoyo District, August 23, 2016.
54. Mbembe, "Necropolitics," 21.
55. Daudi interview.

5. NEGOTIATING LIMINALITY

1. *Harakati* means struggle or movement.
2. The worst thing one might say to a greeting, for instance, is *sawa tu*—just okay.
3. Field notes, September 21, 2016.
4. The old man's choice of the word *kuchinja* ("slaughter / cut throat") is reminiscent of similar metaphors Swahili-speakers had used to describe the ruthless and rapacious nature of colonial extraction. White, *Speaking with Vampires*.
5. On moral economies and everyday forms of resistance see Scott, *Moral Economy of the Peasant*; Scott, *Weapons of the Weak*; Kerkvliet and Scott, *Everyday Forms of Peasant Resistance*; Scott, *Domination and the Arts of Resistance*.
6. Abu-Lughod, "Romance of Resistance"; Ortner, "Resistance and the Problem of Ethnographic Refusal"; Bayat, "From 'Dangerous Classes' to 'Quiet Rebels.'"
7. Scott, *Weapons of the Weak*, 282. On resistance as adaptation see also Carney and Watts, "Manufacturing Dissent."
8. For an excellent example of the Landless Workers Movement (Movimento dos Trabalhadores Rurais Sem Terra; MST) and the multiple subjectivities within the movement see Wolford, *This Land Is Ours Now*.
9. White, *Speaking with Vampires*, 55.
10. In Black America, bell hooks refers to these "homeplaces" as a site of resistance and liberal struggle, especially for Black woman. See hooks, *Yearning: Race, Gender, and Cultural Politics*.
11. Scott, *Domination and the Arts of Resistance*, 25. "Hidden transcripts" can de declared in public, such as in the meeting with the MP, as illustrated in the vignette.
12. Kapferer, *Rumors*, 3.
13. Shibutani, *Improvised News*; White, *Speaking with Vampires*; Kapferer, *Rumors*; Wilson, *Folklore, Gender, and AIDS*.
14. Field notes, October 2015. A literal translation of the proverb is "A person who comes without saying "May I come in?" leaves without saying goodbye."
15. Cohen, *History in Three Keys*, 161.
16. Field notes, March 24, 2016.
17. On discussion of plausibility and rumor see Wilson, *Folklore, Gender, and AIDS*.
18. Malkki, "National Geographic," 25.
19. *Citizen*, "Project Axed to Save Wildlife."
20. Male farmer, interview with the author, Makaani, Bagamoyo District, May 2, 2016.
21. There are many words in Swahili that can be loosely translated as "gossip," "chit-chat," or "light conversation," such as *soga, porojo,* and *mazumgumzo*. To gossip maliciously about someone is *-kumsengenya mtu*.
22. Gluckman, "Gossip and Scandal"; White, *Speaking with Vampires*; Pietilä, *Gossip, Markets, and Gender*; Scott, *Domination and the Arts of Resistance*.
23. Field notes, various dates, 2013–2016.
24. Inbody, "Political Role of Gossip."
25. Inbody; Rysman, "How the 'Gossip' Became a Woman"; Federici, *Caliban and the Witch*.

26. Field notes, March 24, 2016.
27. Field notes, various dates, 2013–2016.
28. Field notes, September 20, 2016.
29. Emmanuel, interview with the author, Razaba, Bagamoyo District, October 4, 2016.
30. For discussion of peasant politics in Africa see Watts, *Silent Violence*; Feierman, *Peasant Intellectuals*; Carney and Watts, "Manufacturing Dissent"; Freidberg, *French Beans and Food Scares*; Schroeder, *Shady Practices*; Neumann, *Imposing Wilderness*; Schroeder, *Shady Practices*.
31. Senior official in the Ministry of Lands, interview with the author, Dar es Salaam, October 21, 2015.
32. Field notes, various dates, 2015–2016. These underground tactics were akin to what scholars and activists have called "guerrilla agriculture" or "guerrilla farming," in which individuals and groups attempt to circumvent laws and regulations that criminalize food production. See Adams and Hardman, "Observing Guerrillas in the Wild"; Cavanagh and Benjaminsen, "Guerrilla Agriculture?"
33. Female elder, interview with the author, No. 5, Bagamoyo District, August 30, 2016.
34. Female farmer, interview with the author, No. 5, Bagamoyo District, August 30, 2016.
35. Male farmer, interview with the author, No. 5, Bagamoyo District, August 30, 2016.
36. Elderly couple, interview with the author, No. 5, Bagamoyo District, August 30, 2016.
37. Female farmer, interview with the author, No. 5, Bagamoyo District, September 7, 2016.
38. Male farmer, interview with the author, No. 5, Bagamoyo District, September 7, 2016.
39. Aziza and Musa, interview with the author, Bagamoyo District, September 7, 2016.
40. Aziza and Musa interview.
41. Saidi, interview with the author, Bagamoyo town, March 18, 2016.
42. Focus group discussion, Mkwaju Mrefu, Bagamoyo District, February 12, 2015.
43. Rukia, interview with the author, Makaani, Bagamoyo District, July 20, 2016.
44. Rukia interview.
45. Focus group discussion, Mkwaju Mrefu, February 12, 2015.
46. Saidi interview.
47. Field notes, August 27, 2016.
48. *Guardian*, "Revealed"; Snyder, "Mothers on the March."
49. Coplan, "You Have Left Me Wandering"; Mbilinyi, "'City' and 'Countryside' in Colonial Tanganyika."
50. Colson and Scudder, *For Prayer and Profit*; Bryceson, "Changing Modalities of Alcohol Usage"; Green, "Trading on Inequality."
51. Neema, interview with the author, Bozi, Bagamoyo District, July 27, 2016.
52. Neema interview.
53. Shabani and Tatu, interview with the author, No. 5, Bagamoyo District, March 18, 2016.
54. Shabani and Tatu interview; field notes, March 18, 2016.
55. Field notes, July 26, 2016.
56. Emmanuel interview.
57. Building on Gramsci's work, Feierman defined peasant intellectuals as those who earned their primary livelihood through farming and who engaged in organizational, directive, and educative activities. See Feierman, *Peasant Intellectuals*.
58. Athumani, interview with the author, Bagamoyo District, June 10, 2016. Athumani's photograph appeared in the introduction.

59. Male farmer, interview with the author, Bozi, Bagamoyo District, July 27, 2016.
60. Male elder, Matipwili, Bagamoyo District, June 10, 2016.
61. On autochthony and citizenship in Africa see Geschiere and Nyamnjoh, "Capitalism and Autochthony"; Geschiere and Jackson, "Autochthony and the Crisis of Citizenship."
62. Field notes, July 26, 2016.
63. Field notes, July 26, 2016.
64. Field notes, August 10, 2016.
65. Field notes, August 10, 2016.
66. Jennifa, interview with the author, No. 4, Bagamoyo District, July 20, 2016.
67. Jennifa, interview with the author, No. 4, Bagamoyo District, September 20, 2016.
68. Female farmer, interview with the author, Bozi, Bagamoyo District, September 20, 2016.
69. Field notes, "Meeting with the wanaharakati," September 30, 2016.
70. Female farmer interview, Bozi, September 20, 2016.
71. Crenshaw, "Mapping the Margins"; Carastathis, *Intersectionality*; Collins and Bilge, *Intersectionality*.

6. OF PRIVILEGE, LAWFARE, AND PERVERSE RESISTANCE

1. Bambadi, interview with the author, Makaani, Bagamoyo District, August 11, 2014.
2. Mwilongo, "Prof Tibaijuka Aamuru"; Machira, "50 Ditch CCM."
3. Bambadi interview.
4. Bambadi interview.
5. *Citizen*, "Villagers Cry over Land Grabbing"; *Citizen*, "Villagers Up in Arms."
6. ActionAid International, "Take Action."
7. Bambadi, interview with the author, Makaani, Bagamoyo District, November 28, 2015.
8. See Cavanagh and Benjaminsen, "Guerrilla Agriculture?"; Grajales, "Land Grabbing, Legal Contention."
9. Japhet and Seaton, *Meru Land Case*; Tenga and Kakoti, "Barabaig Land Case"; Lane, *Pastures Lost*.
10. Ngcukaitobi, *Land Is Ours*.
11. Examples include legal cases against sugarcane and oil palm plantations in Kenya, Cambodia, Uganda, and Guatemala, and a hunting concession in Tanzania. See Neville, "Contentious Political Economy of Biofuels"; IDI, "Case Brief"; Friends of the Earth Europe, "Ugandan Oil Palm Conglomerate"; Oakland Institute, "Losing the Serengeti."
12. Hall et al., "Resistance, Acquiescence or Incorporation?"
13. Alonso-Fradejas, "Anything but a Story Foretold."
14. Ortner, "Resistance and the Problem of Ethnographic Refusal," 179; Abu-Lughod, "Romance of Resistance," 42.
15. In their examination of the relationship between law and dis/order in postcolonial states, Jean and John Comaroff define lawfare as "the resort to legal instruments, to the violence inherent in the law, to commit acts of political coercion, even erasure." The concept initially denoted efforts by the colonial state to conquer and control indigenous populations through the use of legal force, including the invention and manipulation of law to legitimize land expropriation and dispossession. With the prevalence of "culture of legality" in the postcolonial world, however, the Comaroffs contend that a wide range of actors, including "the weak, the strong, and everyone in between" can now appropriate legal instruments. See Comaroff and Comaroff, *Law and Disorder*, 306, 30; Colson, "Impact of the Colonial Period"; Ranger, "Invention of Tradition"; Chanock, *Law, Custom, and Social Order*; Tenga, "Legitimizing Dispossession"; Comaroff, "Symposium Introduction"; Alden Wily, "Looking Back to See Forward."

16. Tenga and Kakoti, "Barabaig Land Case"; Hughes, *Moving the Maasai*; Bowd, "Access to Justice in Africa"; Askew, Maganga, and Odgaard, "Of Land and Legitimacy"; Cavanagh and Benjaminsen, "Guerrilla Agriculture?"; Dancer, "An Equal Right to Inherit?"; Logan, "Ambitious SDG Goal"; Gilbert, "Strategic Litigation Impacts."

17. Salum Yusuf, Ally Thabiti, Ally Said and 537 Others v. Agro EcoEnergy Company Limited, Bagamoyo District Council, Bagamoyo EcoEnergy Limited, the Commissioner for Lands, the Attorney General, High Court of Tanzania (Land Division) at Dar es Salaam, Land Case No. 162 of 2012. I use real names of all parties involved since the case is public record.

18. URT, Village Land Act, 1999, Part 4, Section 57(1).

19. Prominent Tanzanian lawyer and legal scholar, interview with the author, Dar es Salaam, March 1, 2016.

20. Agro EcoEnergy Tanzania, "White Paper."

21. Per Carstedt, interview with the author, Dar es Salaam, August 22, 2014.

22. Africa Confidential, "Sour Fate of Sugar Project."

23. Field notes, February 7, 2016.

24. Male farmer, interview with the author, Matipwili, Bagamoyo District, July 30, 2016.

25. Male farmer, interview with the author, Gama, Bagamoyo District, August 6, 2014.

26. Male farmer, interview with the author, Makaani, Bagamoyo District, March 15, 2016.

27. Female farmer, interview with the author, Matipwili, Bagamoyo District, September 4, 2016.

28. Male migrant farmer, interview with the author, Makaani, Bagamoyo District, August 23, 2016.

29. Field notes, February 7, 2016.

30. Hadija, interview with the author, Makaani, Bagamoyo District, February 12, 2016.

31. Minutes of the meeting of Kisauke subvillage of Matipwili, following the transfer of their land to TANAPA (Saadani), November 30, 2003.

32. Letter from Ally Thabiti Ngwega to Makurunge village chairman, dated November 30, 2003.

33. Thabiti, interview with the author, Makaani, Bagamoyo District, February 11, 2015.

34. Letter from "Gama Kitame Sub-village" to Makurunge village executive officer, signed by Ally Thabiti Ngwega, dated January 5, 2010.

35. Letter from Makaani (Gama) Residents to the Bagamoyo District Commissioner, dated March 10, 2010.

36. Letter from RAZABA manager to Bagamoyo district commissioner, dated November 15, 2010.

37. Tanzania Police Force charge sheet, dated December 20, 2010.

38. Anonymous source, interview with the author, Bagamoyo town, November 16, 2015.

39. Thabiti interview.

40. Fatuma, interview with the author, Makaani, Bagamoyo District, March 9, 2016.

41. Fatuma interview.

42. Fatuma interview.

43. Hodgson, "'My Daughter'"; Tsikata, "Securing Women's Interests"; TAWLA, "Position Paper on Gender"; Dancer, *Women, Land and Justice*.

44. Male farmer, interview with the author, Bozi, Bagamoyo District, August 19, 2016.

45. Female farmer, interview with the author, Makaani, Bagamoyo District, February 10, 2016.

46. Hadija interview.

47. Hadija interview.
48. Fatuma interview.
49. Male elder, interview with the author, Gama, Bagamoyo District, June 18, 2016.
50. Village chairman, interview with the author, Matipwili, Bagamoyo District, June 10, 2016.
51. Subvillage chairman, interview with the author, Kitame, Bagamoyo District, September 3, 2016.
52. Comaroff and Comaroff, *Law and Disorder*, 32.
53. Similar observations have been made by Askew, Maganga, and Odgaard, "Of Land and Legitimacy"; Bluwstein et al., "Between Dependence and Deprivation."

CONCLUSION

1. Mbashiru, "Bakhresa Lands Sugar Covenant."
2. Mbashiru, "Bakhresa Lands Sugar Covenant."
3. Mbashiru, "Bakhresa Lands Sugar Covenant." The same month the president announced his land gift to Bakhresa, he reportedly transferred over twenty-five thousand hectares in Morogoro District to national retirement schemes—the National Social Security Fund and the Parastatal Pension Fund—so they could also invest in sugarcane production toward national sugar self-sufficiency. See Omar, "NSSF, PPF Link Up"; *Guardian*, "Mkulazi Sugar Project."
4. One of Said Salim Bakhresa's sons is the managing director of Bagamoyo Sugar Limited, and a businessman from India with experience operating sugar factories there is its chief executive officer.
5. Lugongo, "Bakhresa Kicks off Sugar Production"; Mwagonde, "Bakhresa Firm to Start"; URT, "Waziri Mkuu." Whereas EcoEnergy planned to mobilize its start-up capital from international financial institutions and development agencies, Bakhresa would do so from national and commercial banks, including the Tanzanian Agricultural Development Bank, the CRDB Bank, and the Standard Chartered Bank.
6. Saidi, interview with the author, Bagamoyo town, June 6, 2016.
7. Saidi interview.
8. Agro EcoEnergy Tanzania, "White Paper." The white paper notes that the government of Tanzania had already revoked the land title as of April 15, 2016.
9. Agro EcoEnergy Tanzania, "White Paper"; Global Arbitration Review, "Tanzania Faces ICSID Claim"; ICSID, "EcoDevelopment in Europe AB."
10. Per Carstedt, email correspondence with the author, September 15, 2017.
11. Global Arbitration Review, "Tanzania Faces ICSID Claim."
12. Agro EcoEnergy Tanzania, "White Paper," 19.
13. Agro EcoEnergy Tanzania, "White Paper," 19.
14. Attorney general quoted in Ng'hily, "Air Tanzania Plane Seized." While the ICSID tribunal has provisionally postponed the enforcement of the award in light of Tanzania's application for annulment, EcoEnergy reportedly persuaded a Dutch court to seize an Air Tanzania plane that was grounded in the Netherlands because of an engine problem. See also Ch-aviation, "Air Tanzania A220-300 Remains Attached."
15. See ICSID, "EcoDevelopment in Europe AB." Under the ICSID Convention, a tribunal's decision is final and binding, but the parties can seek post-award remedies, which can take the case in new and unpredictable directions. As Heiskanen and Halonen ("Post-award Remedies") explain, the overall length of the case may become excessive as one party's successful application for annulment can lead to the other party's resubmission of the case, followed by a second, third, or fourth set of annulment proceedings.
16. Berry, *No Condition Is Permanent*, 43–66.

17. The nickname is also attributed to Magufuli's track record of promoting large-scale infrastructure projects during his tenure as minister of works, transport, and communications under the Mkapa and Kikwete administrations (2000–2005 and 2010–2015 respectively).

18. Under the Kikwete administration, the government generally held a favorable view toward foreign direct investments (FDIs), and Tanzania was named one of the top destinations for investors in Africa in 2014 by the UN Conference on Trade and Development. Since Magufuli's entering office, FDI levels dropped to half those under his predecessor's regime. See US Department of State, "2020 Investment Climate Statements: Tanzania."

19. Policy Forum, "Tanzania Governance Review," 2015–16.

20. Ogola, "#Whatwouldmagufulido?"; Mohammed, "Twitter Is Having Fun"; Paget, "Again, Making Tanzania Great."

21. Makoye, "Tanzania Seizes 'Idle' Land."

22. On resource nationalism under the Kikwete and Magufuli administrations see Jacob and Pedersen, "New Resource Nationalism?"

23. These laws and regulations include the Natural Wealth and Resources (Permanent Sovereignty) Bill, the Natural Wealth and Resources Contracts (Review and Re-negotiation of Unconscionable Terms) Bill, the Written Laws (Miscellaneous Amendments) Act, and the Public-Private Partnership Act (Amendment). In late 2018, the government also terminated its bilateral investment treaty with the Netherlands, contending that international arbitration institutions lack impartiality and are biased toward investor interests. The passing of the Arbitration Act in February 2020, which replaced the colonial Arbitration Ordinance of 1932, further restricted the use of arbitration for resolving investment disputes beyond the resources and energy sector. See Masamba, "Government Regulatory Space"; Ballantyne, "Tanzania Proposes Arbitration Reforms"; Karashani, "Tanzania Reviews Law on Mediation"; Amir, "Wind of Change!"; Kdanka, "Tanzania Ends Investment Treaty"; Muchari, "Tanzania Cites Bias."

24. Rweyemamu, "Cost of JPM's Economic Warfare."

25. *Citizen*, "Magufuli: We'll Negotiate on Our Own Terms."

26. Dar24 Media, "Breaking News," 24.

27. Brennan, "Blood Enemies."

28. On "organized forgetting" in postsocialist Africa see Pitcher, "Forgetting from Above"; Schroeder, *Africa after Apartheid*.

29. While the two corporate investors, EcoEnergy and Bakhresa, were distinct in their race, ethnicity, and nationality, people perceived them as similar by virtue of their privileged class position. As one displaced male farmer in Matipwili put it, "The only difference between them is that one is white [*mzungu*] and the other is Asian [*mhindi*]. And we are the natives [*wenyeji*]. Whoever the investors are, they are greedy; they are always benefiting and will always benefit more than the natives" (Male elder, interview with the author, Matipwili, Bagamoyo District, October 5, 2020). On Magufuli's nostalgia's for Nyerere's Tanzania see Kanyabwoya, "Magufuli's Method in 'Madness'"; Paget, "Again, Making Tanzania Great."

30. State House, "Rais Dkt John Magufuli," emphasis added.

31. *Citizen*, "Lissu Charged with Sedition"; Amnesty International, "Price We Pay"; Human Rights Watch, "Tanzania: Freedoms Threatened"; Article 19, "Shrinking Civic Space."

32. Lissu, "Tanzania's Five Years of Devastation."

33. *BBC News*, "Coronavirus"; Bariyo and Parkinson, "Tanzania's Leader Urges."

34. OHCHR, "In Her Global Human Rights Update."

35. Dahir, "As Tanzania Votes."

36. *BBC News*, "Tanzania Elections."
37. Burke, "Tanzanian Opposition Accuses Police."
38. *Guardian*, "Police Sound Warning."
39. Anna, "Tanzanian Opposition"; Burke, "Tanzanian Government Cracks Down."
40. Burke, "Tanzania's Covid-Denying President."
41. Lissu, "Tanzania's Five Years of Devastation."
42. Owere, "President Samia."
43. Bariyo, "Tanzania's New President"; US Department of State, "2021 Investment Climate Statements"; Habari Leo, "PM Assures Investors"; Schipani, "Tanzania's New President."
44. Askew, "Sung and Unsung," 28.
45. As this book goes to press, Bakhresa has begun commercial production on about twelve hundred hectares. News reports from October 2022 suggest that the company's plan for expanding production has been frustrated by problems with water access. See Nachilongo, "Water Scarcity Blow to Bakhresa."
46. *Citizen*, "What Holds Up Sh660bn Bagamoyo Sugar Project."
47. *Citizen*, "What Holds Up Sh660bn Bagamoyo Sugar Project."
48. Letter from Matipwili villagers to the prime minister, dated April 18, 2020, translated by the author. The original letter read: "... tunapoondolewa katika Makazi na Mashamba yetu tukakae wapi? Tulime wapi na tutaishi *vipi*, tutawasomeshaje watoto wetu na shamba ni kila kilu katika maisha yetu. LENGO: Tupatiwe Makazi na Mashamba na baadae ndipo tuondoke na malipo kuangaliwa upya haki haikutendeka katika malipo. Malipo hayakuwa wazi hakuna anaetambua kalipwa nini. Mwisho; Tunaomba Mkuu wa Mkoa utuangalie kwa jicho la tatu. Tunakutakia kazi njema katika Ujenzi wa Taifa." The penultimate sentence, "utuangalie kwa jicho la tatu" (literally, look at us with a third eye), can be translated as the villagers requesting the commissioner to take a neutral, third-party perspective, but it could also be interpreted as the villagers asking for empathy and compassion.
49. Azam TV, "Mkuu Wa Mkoa Pwani."
50. Azam TV, "Mkuu Wa Mkoa Pwani."
51. Chief valuer, Ministry of Lands, Housing, and Human Settlements Development, interview with the author, Dar es Salaam, August 18, 2016.
52. On the relationship between time / temporal politics and bureaucratic state power see Gonçalves, "Orientações Superiores."
53. Rwegasira, *Land as a Human Right*; Kironde, "Improving Land Sector Governance"; Oxfam Tanzania, "Balancing Infrastructure Development"; Kusiluka et al., "Negative Impact"; Kombe, "Land Acquisitions for Public Use"; Pedersen and Kweka, "Political Economy of Petroleum Investments."
54. Rama, interview with the author, Kibuyu Mimba, Bagamoyo District, October 22, 2020.
55. On "liminal collectivity" see Malkki, *Purity and Exile*.
56. Suba, interview with the author, Matipwili, Bagamoyo District, October 12, 2020.
57. Displaced residents, interviews with the author, Bagamoyo District, various dates, October 2020.
58. Displaced residents interviews.
59. Barabaig youth leader, interview with the author, Bagamoyo District, October 8, 2020.
60. It is unclear how Bambadi was able to amass such a disproportionate amount of compensation, though one might speculate on his relationship with sugar importers, possibly including Bakhresa, which EcoEnergy alleged had sponsored the court case.
61. Rama interview.

62. Mwajuma, interview with the author, Matipwili, Bagamoyo District, October 12, 2020.

63. Mohammedi, interview with the author, Matipwili, Bagamoyo District, October 12, 2020; Mwanahamisi, interview with the author, Matipwili, Bagamoyo District, October 12, 2020; displaced residents interviews.

64. Rama interview.

65. Rama interview. Grappling with the disjuncture between his aspirations, cultural gender norms, and sense of self, Rama felt shame in falling short of what he thought a respectable man in rural Tanzania ought to be able to do: "provide for his family." As noted in chapter 3, a similar theme appears in the literature on the shifting meanings of masculinity, especially among unemployed male youth across Africa. See, for example, Cornwall, "To Be A Man"; Lindsay and Miescher, *Men and Masculinities*.

66. Barabaig youth leader interview. Though not discussed in depth in this book, farmer-herder conflicts have historically persisted in Tanzania and across Africa. See, for example, Benjaminsen, Maganga, and Abdallah, "Kilosa Killings"; Walwa, "Land Use Plans in Tanzania"; Krätli and Toulmin, *Farmer-Herder Conflict*.

67. Suba interview.

68. Mwajuma interview.

69. URT, *Annual General Report*, 140–41; *Guardian*, "CAG."

70. For possible trajectories see Mbilinyi, "Sweet and Sour"; Norris and Worby, "Sexual Economy"; Sulle and Dancer, "Gender, Politics, and Sugarcane."

71. ICSID, "Cases Database"; UNCTAD, "Investment Policy Hub."

72. According to a recent OECD study that surveyed a sample of 1,660 BITs, over 93 percent of the treaties provided access to international arbitration exclusively or in addition to domestic judicial reviews. See OECD, "Dispute Settlement Provisions."

73. Between 1972 and 1996, ICSID was registering on average 1.5 cases per year, with a total caseload of thirty-eight over this twenty-five-year period. And as of December 2019, the cumulative number of BIT-based arbitration cases submitted to ICSID and non-ICSID bodies had reached over one thousand. The figure includes cases categorized under "oil, gas, and mining," "electric power and other energy," "agriculture, fishing and forestry," and "water, sanitation, and flood protection." See TNI, "Transatlantic Corporate Bill of Rights"; ICSID, "ICSID Caseload Statistics"; UNCTAD, "Fact Sheet."

74. Peterson, "Land Deals Could Sow."

75. ICSID, "Standard Chartered Bank."

76. UNCTAD, "Nachingwea and Others."

77. For context, Tanzania's current national budget for the 2020/21 fiscal year is USD 15 billion, 31 percent of which is to be financed through grants and concessional loans from development agencies, as well as nonconcessional loans from domestic and foreign sources. The nation's total debt as of April 2020 stands at USD 24 billion. See Namkwahe, "Tanzania's National Debt"; URT, Speech by the Minister of Finance and Planning.

78. TNI, "License to Grab"; Eberhardt and Olivet, "Profiting from Injustice"; Borras et al., "Transnational Land Investment Web"; Cotula, "Land Rights and Investment Treaties"; Cotula and Schröder, "Community Perspectives."

79. Cotula and Schröder, "Community Perspectives."

80. Cotula and Schröder, "Community Perspectives."

81. To disembed the economy from society, or to allow the market to be the sole director of life, Polanyi reminds us, "would result in the demolition of society." Polanyi, *Great Transformation*, 76.

82. TNI, "License to Grab."

83. For similar critiques see Collins, "Governing the Global Land Grab."

84. For these reasons, Borras and Franco ("From Threat to Opportunity?," 15) argue that these voluntary codes of conduct "should not be considered, even as a second-best approach."
85. De Schutter, "How Not to Think of Land-Grabbing," 275.
86. Weil, *Need for Roots*, 83.
87. Tuck, "Suspending Damage," 416.
88. hooks, *Yearning: Race, Gender, and Cultural Politics*, 57.
89. Mwanahamisi, interview with the author, Gama/Matipwili, Bagamoyo District, June 4, 2016.

Bibliography

Abdallah, Jumanne, Linda Engström, Kjell Havnevik, and Lennart Salomonsson. "Large-Scale Land Acquisitions in Tanzania: A Critical Analysis of Practices and Dynamics." In *The Global Land Grab: Beyond the Hype*, edited by Mayke Kaag and Annelies Zoomers, 46–53. London: Zed Books, 2014.

Abrams, Philip. "Notes on the Difficulty of Studying the State (1977)." *Journal of Historical Sociology* 1, no. 1 (1988): 58–89.

Abu-Lughod, Lila. "The Romance of Resistance: Tracing Transformations of Power through Bedouin Women." *American Ethnologist* 17, no. 1 (1990): 41–55.

ActionAid International. *Take Action: Stop EcoEnergy's Land Grab in Bagamoyo, Tanzania*. Johannesburg: ActionAid International, 2015.

ActionAid Sweden. "Sekab—Ethanol at Any Cost?" Stockholm: ActionAid Sweden, 2009.

ActionAid Tanzania. "Implication of Biofuels Production on Food Security in Tanzania." Dar es Salaam: ActionAid Tanzania, 2009.

Adams, David, and Michael Hardman. "Observing Guerrillas in the Wild: Reinterpreting Practices of Urban Guerrilla Gardening." *Urban Studies* 51, no. 6 (May 1, 2014): 1103–19.

AfDB. "African Development Bank Group's Integrated Safeguards System: Policy Statement and Operational Safeguards." Tunis: AfDB, 2013.

———. "Bagamoyo Sugar Project, Tanzania. Executive Summary of the Resettlement Action Plan." Tunis: AfDB, 2012.

———. "Involuntary Resettlement Policy." Tunis: African Development Bank, 2003.

Africa Confidential. "Sour Fate of Sugar Project." *Africa Confidential* 58, no. 5 (March 3, 2017): 10–11.

Agarwal, Bina. "'Bargaining' and Gender Relations: Within and beyond the Household." *Feminist Economics* 3, no. 1 (1997): 1–51.

Agro EcoEnergy Tanzania. "Bagamoyo EcoEnergy Community and Outgrower Development Programme." Dar es Salaam: Agro EcoEnergy Tanzania, September 2012.

———. "Draft Resettlement Action Plan for the EcoEnergy Sugar Cane Plantation and Processing Plant." Dar es Salaam: Agro EcoEnergy Tanzania, 2012.

———. "Early Works Launch of the Agro EcoEnergy Sugar Project in Bagamoyo, 15 March 2014." 2014. https://www.youtube.com/watch?v=327Krd5jd2s.

———. "Press Release on the Bagamoyo EcoEnergy Project in Bagamoyo, Coast Region, Tanzania." March 23, 2015.

———. "Response to ActionAid Report." Dar es Salaam: Agro EcoEnergy Tanzania, 2015.

———. "White Paper on the Bagamoyo EcoEnergy Project in Tanzania." Dar es Salaam: Agro EcoEnergy Tanzania, 2017.

Akram-Lodhi, A. Haroon, and Cristóbal Kay. *Peasants and Globalization: Political Economy, Rural Transformation and the Agrarian Question*. New York: Routledge, 2009.

Alden Wily, Liz. "Compulsory Acquisition as a Constitutional Matter: The Case in Africa." *Journal of African Law* 62, no. 1 (February 2018): 77–103.

———. "Looking Back to See Forward: The Legal Niceties of Land Theft in Land Rushes." *Journal of Peasant Studies* 39, no. 3–4 (2012): 751–75.

Alonso-Fradejas, Alberto. "Anything but a Story Foretold: Multiple Politics of Resistance to the Agrarian Extractivist Project in Guatemala." *Journal of Peasant Studies* 42, no. 3–4 (2015): 489–515.

Alpers, Edward A. *Ivory and Slaves: Changing Pattern of International Trade in East Central Africa to the Later Nineteenth Century*. Berkeley: University of California Press, 1975.

Aminzade, Ronald. "The Dialectic of Nation Building in Postcolonial Tanzania." *Sociological Quarterly* 54, no. 3 (August 2013): 335–66.

———. *Race, Nation, and Citizenship in Post-colonial Africa: The Case of Tanzania*. Cambridge: Cambridge University Press, 2013.

Amir, Ibrahim. "A Wind of Change! Tanzania's Attitude towards Foreign Investors and International Arbitration." *Kluwer Arbitration Blog*, December 28, 2018.

Amnesty International. "The Price We Pay: Targeted for Dissent by the Tanzanian State." London: Amnesty International, 2019.

Anderson, Benedict. *Imagined Communities: Reflections on the Origin and Spread of Nationalism*. London: Verso, 2006.

Anna, Cara. "Tanzanian Opposition: Colleagues Face Terror-Related Charges." *Washington Post*, 2020. https://www.washingtonpost.com/world/africa/tanzanias-opposition-says-police-blocking-protest-over-vote/2020/11/02/5f3e8c52-1cde-11eb-ad53-4c1fda49907d_story.html.

Anzaldúa, Gloria E. *Borderlands / La Frontera: The New Mestiza*. San Francisco: Aunt Lute Books, 1987.

———. "(Un)Natural Bridges, (Un)Safe Spaces." In *This Bridge We Call Home: Radical Visions for Transformation*, edited by Gloria E. Anzaldúa and AnaLouise Keating, 1–5. New York: Routledge, 2002.

Arezki, Rabah, Klaus Deininger, and Harris Selod. "What Drives the Global 'Land Rush'?" Washington, DC: World Bank, 2011.

Article 19. "Shrinking Civic Space ahead of the October 2020 Elections." September 11, 2020. https://www.article19.org/resources/tanzania-shrinking-civic-space/.

Ashura Yussuf. "Speech on New Land Policy in Tanzania." 2012. https://www.youtube.com/watch?v=x_oaLqKzEnw&t=2s.

Askew, Kelly. "Sung and Unsung: Musical Reflections on Tanzanian Postsocialisms." *Africa: Journal of the International African Institute* 76, no. 1 (2006): 15–43.

Askew, Kelly, Faustin Maganga, and Rie Odgaard. "Of Land and Legitimacy: A Tale of Two Lawsuits." *Africa* 83, no. 1 (2013): 120–41.

Askew, Kelly, and Rie Odgaard. "Deeds and Misdeeds: Land Titling and Women's Rights in Tanzania." *New Left Review* 118 (July/August 2019): 68–85.

Auyero, Javier. *Patients of the State: The Politics of Waiting in Argentina*. Durham, NC: Duke University Press, 2012.

Azam TV. "Kiwanda Cha Bagamoyo Sugar Kukamilika Hivi Karibuni." May 26, 2021. https://www.youtube.com/watch?v=6aWu9F7s0v4.

———. "Mkuu Wa Mkoa Pwani Aamuru Wavamizi Shamba La Miwa La Bakhresa Kuondoka Haraka." August 16, 2018. https://www.youtube.com/watch?v=8MOJ9h-RFqA.

Bakhresa. "Company Profile." Azam Bakhresa Group, n.d. http://bakhresa.com/about-us/.

Baldus, Rolf D., K. Roettcher, and D. Broska. "Saadani: An Introduction to Tanzania's Future 13th National Park." Dar es Salaam: Wildlife Division, Deutsche Gesellschaft für Technische Zusammenarbeit, GTZ Wildlife Programme in Tanzania, 2001.

Ballantyne, Jack. "Tanzania Proposes Arbitration Reforms amid Investor Backlash." Global Arbitration Review, February 12, 2020.

Bariyo, Nicholas. "Tanzania's New President Nudges Country Away from Covid-19 Denial." Wall Street Journal, April 12, 2021, sec. World. https://www.wsj.com/articles/tanzanias-president-nudges-country-away-from-covid-19-denial-11618247113.

Bariyo, Nicholas, and Joe Parkinson. "Tanzania's Leader Urges People to Worship in Throngs against Coronavirus." Wall Street Journal, April 8, 2020. https://www.wsj.com/articles/tanzanias-leader-urges-people-to-worship-in-throngs-against-coronavirus-11586347200.

Bayat, Asef. "From 'Dangerous Classes' to 'Quiet Rebels': Politics of the Urban Subaltern in the Global South." *International Sociology* 15, no. 3 (September 1, 2000): 533–57.

BBC News. "Coronavirus: John Magufuli Declares Tanzania Free of Covid-19." June 8, 2020, sec. Africa. https://www.bbc.com/news/world-africa-52966016.

———. "Tanzania Elections: President Magufuli in Landslide Win amid Fraud Claims." October 30, 2020, sec. Africa. https://www.bbc.com/news/world-africa-54748332.

———. "Indian Ocean Dipole: What Is It and Why Is It Linked to Floods and Bushfires?" December 7, 2019, sec. Science & Environment. https://www.bbc.com/news/science-environment-50602971.

Beban, Alice. *Unwritten Rule: State-Making through Land Reform in Cambodia*. Ithaca, NY: Cornell University Press, 2021.

Beidelman, Thomas O. *The Matrilineal Peoples of Eastern Tanzania (Zaramo, Luguru, Kaguru, Ngulu, Etc.)*. London: International African Institute, 1967.

Benjaminsen, Tor A., Faustin Maganga, and Jumanne Abdallah. "The Kilosa Killings: Political Ecology of a Farmer–Herder Conflict in Tanzania." *Development and Change* 40, no. 3 (2009): 423–45.

Berry, Sara. *No Condition Is Permanent: The Social Dynamics of Agrarian Change in Sub-Saharan Africa*. Madison: University of Wisconsin Press, 1993.

Bhabha, Homi K. *The Location of Culture*. London: Routledge, 1994.

Biovision. "Can Agri-business Really Improve Food Security and the Livelihood of Smallholders in Tanzania?" October 7, 2015. https://www.biovision.ch/en/news/sagcot/.

Bluwstein, Jevgeniy, Jens Friis Lund, Kelly Askew, Howard Stein, Christine Noe, Rie Odgaard, Faustin Maganga, and Linda Engström. "Between Dependence and Deprivation: The Interlocking Nature of Land Alienation in Tanzania." *Journal of Agrarian Change* 18, no. 4 (2018): 806–30.

Booth, Frances E., and G. Wickens. "Non-timber Uses of Selected Arid Zone Trees and Shrubs in Africa." FAO Conservation Guide. Rome: FAO, 1988.

Booth, Vernon. "Options and the Future Status and Use of Saadani/Mkwaja Game Reserve." Dar es Salaam: Saadani Conservation and Development Programme, 2000.

Borras, Saturnino M., Jr., and Jennifer C. Franco. "From Threat to Opportunity? Problems with the Idea of a 'Code of Conduct' for Land-Grabbing." *Yale Human Rights and Development Law Journal* 13 (2010): 507–23.

Borras, Saturnino M., Jr., Jennifer C. Franco, S. Ryan Isakson, Les Levidow, and Pietje Vervest. "The Rise of Flex Crops and Commodities: Implications for Research." *Journal of Peasant Studies* 43, no. 1 (2016): 93–115.

Borras, Saturnino M., Jr., Elyse N. Mills, Philip Seufert, Stephan Backes, Daniel Fyfe, Roman Herre, and Laura Michéle. "Transnational Land Investment Web: Land Grabs, TNCs, and the Challenge of Global Governance." *Globalizations* 17, no. 4 (2019): 608–28.

Bowd, Richard. "Access to Justice in Africa: Comparisons between Sierra Leone, Tanzania and Zambia." Pretoria: Institute for Security Studies, 2009.

Brennan, James R. "Blood Enemies: Exploitation and Urban Citizenship in the Nationalist Political Thought of Tanzania, 1958–75." *Journal of African History* 47, no. 3 (2006): 389–413.

———. *Taifa: Making Nation and Race in Urban Tanzania*. Athens: Ohio University Press, 2012.

———. "Youth, the TANU Youth League and Managed Vigilantism in Dar Es Salaam, Tanzania, 1925–73." *Africa* 76, no. 2 (2006): 221–46.

Brown, Walter T. "Bagamoyo: An Historical Introduction." *Tanzania Notes and Records* 71, no. 1970 (1970): 69–83.

———. "Bagamoyo: Inquiry into an East African Place Name." In *Place Names in Africa: Colonial Urban Legacies, Entangled Histories*, edited by Liora Bigon, 37–44. Geneva: Springer, 2016.

———. "The Politics of Business: Relations between Zanzibar and Bagamoyo in the Late Nineteenth Century." *African Historical Studies* 4, no. 3 (1971): 631–43.

———. "A Pre-colonial History of Bagamoyo: Aspects of the Growth of an East African Coastal Town." PhD diss., Boston University, 1971.

Brown, Wendy. *States of Injury: Power and Freedom in Late Modernity*. Princeton, NJ: Princeton University Press, 1995.

Bryceson, Deborah F. "Changing Modalities of Alcohol Usage." In *Alcohol in Africa: Mixing Business, Pleasure and Politics*, edited by Deborah Bryceson, 22–52. Portsmouth, NH: Heinemann, 2002.

———. "Gender Relations in Rural Tanzania: Power Poetics or Cultural Consensus?" In *Gender, Family and Household in Tanzania*, edited by Colin Creighton and Cuthbert K. Omari, 37–69. Aldershot, UK: Avebury, 1995.

Burke, Jason. "Tanzanian Government Cracks Down on Opposition after Disputed Election." *Guardian*, November 2, 2020. http://www.theguardian.com/world/2020/nov/02/tanzanian-opposition-figures-arrested-after-disputed-election.

———. "Tanzanian Opposition Accuses Police of Killing Nine during Protests." *Guardian*, October 27, 2020. https://www.theguardian.com/world/2020/oct/27/tanzanian-president-accused-of-repression-on-eve-of-election.

———. "Tanzania's Covid-Denying President, John Magufuli, Dies Aged 61." *Guardian*, March 18, 2021. http://www.theguardian.com/world/2021/mar/17/tanzanias-president-john-magufuli-dies-aged-61.

Burnod, Perrine, Mathilde Gingembre, and Rivo Andrianirina Ratsialonana. "Competition over Authority and Access: International Land Deals in Madagascar." *Development and Change* 44, no. 2 (2013): 357–79.

Butler, Judith. *Gender Trouble: Feminism and the Subversion of Identity*. New York: Routledge, 1999.

Byamugisha, Frank F. K. "Securing Africa's Land for Shared Prosperity: A Program to Scale up Reforms and Investments." Washington, DC: World Bank, 2013.

Calvert, Albert Frederick. *German East Africa*. London: T. W. Laurie, 1917.

Carastathis, Anna. *Intersectionality: Origins, Contestations, Horizons*. Lincoln: University of Nebraska Press, 2016.
Carmody, Pádraig. *The New Scramble for Africa*. Cambridge: Polity, 2011.
Carney, Judith. "Gender Conflict in Gambian Wetlands." In *Liberation Ecologies: Environment, Development, Social Movements*, edited by Richard Peet and M. Watts, 2nd ed., 289–308. London: Routledge, 2004.
Carney, Judith, and Michael Watts. "Manufacturing Dissent: Work, Gender and the Politics of Meaning in a Peasant Society." *Africa: Journal of the International African Institute* 60, no. 2 (1990): 207–41.
Carstedt, Per. "Sustainable Ethanol: What Is the Context?" Presentation at the International Energy Agency Bioenergy executive committee workshop, Oslo, Norway, May 14, 2008. https://www.ieabioenergy.com/wp-content/uploads/2013/10/P04-Sustainable-Ethanol-What-is-the-Context-P.-Carstedt.pdf.
Cavanagh, Connor Joseph, and Tor A. Benjaminsen. "Guerrilla Agriculture? A Biopolitical Guide to Illicit Cultivation within an IUCN Category II Protected Area." *Journal of Peasant Studies* 42, no. 3–4 (2015): 725–45.
Cernea, Michael, and Christopher McDowell. *Risks and Reconstruction: Experiences of Resettlers and Refugees*. Washington, DC: World Bank, 2000.
Chachage, Chambi. "Land Acquisition and Accumulation in Tanzania: The Case of Morogoro, Iringa, and Pwani Regions." Dar es Salaam: PELUM Tanzania, 2010.
Chanock, Martin. *Law, Custom, and Social Order: The Colonial Experience in Malawi and Zambia*. Portsmouth, NH: Heinemann, 1985.
Chase, Hank. "The Zanzibar Treason Trial." *Review of African Political Economy* 6 (May–August 1976): 14–33.
Ch-aviation. "Air Tanzania A220-300 Remains Attached over Land Use Claim." ch-aviation, December 6, 2022. https://www.ch-aviation.com/portal/news/122174-air-tanzania-a220-300-remains-attached-over-land-use-claim.
Chung, Youjin B. "Engendering the New Enclosures: Development, Involuntary Resettlement and the Struggles for Social Reproduction in Coastal Tanzania." *Development and Change* 48, no. 1 (2017): 98–120.
———. "Governing a Liminal Land Deal: The Biopolitics and Necropolitics of Gender." *Antipode* 52, no. 3 (2020): 722–41.
———. "The Grass Beneath: Conservation, Agro-industrialization, and Land-Water Enclosures in Postcolonial Tanzania." *Annals of the American Association of Geographers* 109, no. 1 (2019): 1–17.
Chung, Youjin B., and Marie Gagné. "Understanding Land Deals in Limbo in Africa: A Focus on Actors, Processes, and Relationships." *African Studies Review* 64, no. 3 (2021): 595–604.
Chung, Youjin B., Sera Lewise Young, and Rachel Bezner Kerr. "Rethinking the Value of Unpaid Care Work: Lessons from Participatory Visual Research in Central Tanzania." *Gender, Place & Culture* 26, no. 11 (2019): 1544–69.
Citizen. "Dar Land Sharks Target Key Deal in Bagamoyo." March 12, 2014. https://www.thecitizen.co.tz/tanzania/news/national/-dar-land-sharks-target-key-deal-in-bagamoyo--2506424.
———. "Lissu Charged with Sedition at Kisutu Court." July 24, 2017. https://www.thecitizen.co.tz/news/Lissu-charged-with-sedition-at-Kisutu-Court/1840340-4030256-158c9ju/index.html.
———. "Magufuli: We'll Negotiate on Our Own Terms." June 13, 2017. https://www.thecitizen.co.tz/News/Magufuli—We-ll-negotiate-on-our-own-terms/1840340-3968092-format-xhtml-pe42i6/index.html.
———. "Project Axed to Save Wildlife." May 20, 2016. https://www.thecitizen.co.tz/tanzania/news/national/state-project-axed-to-save-wildlife--2555734.

———. "Unanswered Questions as BRN Disbanded." June 28, 2017. https://www.the citizen.co.tz/tanzania/news/national/unanswered-questions-as-brn-disbanded-2594458.
———. "Villagers Cry over Land Grabbing." March 13, 2013. https://www.farmland grab.org/post/view/21776-tanzania-villagers-cry-over-land-grabbing.
———. "Villagers up in Arms for Fear of Losing Land." January 29, 2014.
———. "What Holds Up Sh660bn Bagamoyo Sugar Project." January 25, 2018. https://www.thecitizen.co.tz/tanzania/magazines/what-holds-up-sh660bn-bagamoyo-sugar-project-2621012.
Cohen, Paul A. *History in Three Keys: The Boxers as Event, Experience, and Myth*. New York: Columbia University Press, 1997.
Collins, Andrea M. "Governing the Global Land Grab: What Role for Gender in the Voluntary Guidelines and the Principles for Responsible Investment?" *Globalizations* 11, no. 2 (2014): 189–203.
Collins, Patricia Hill, and Sirma Bilge. *Intersectionality*. Cambridge: Polity, 2016.
Colson, Elizabeth. "Gendering Those Uprooted by 'Development.'" In *Engendering Forced Migration: Theory and Practice*, edited by Doreen Marie Indra, 23–39. New York: Berghahn Books, 1999.
———. "The Impact of the Colonial Period on the Definition of Land Rights." In *Colonialism in Africa, 1870–1960*. Vol. 3, *Profiles of Change: African Society and Colonial Rule*, edited by Victor Turner. Cambridge: Cambridge University Press, 1971.
———. *The Social Consequences of Resettlement: The Impact of the Kariba Resettlement upon the Gwembe Tonga*. Manchester: Manchester University Press, 1971.
Colson, Elizabeth, and Thayer Scudder. *For Prayer and Profit: The Ritual Economic and Social Importance of Beer in Gwembe District, Zambia, 1950–1982*. Stanford, CA: Stanford University Press, 1988.
Comaroff, Jean, and John L. Comaroff. *Law and Disorder in the Postcolony*. Chicago: University of Chicago Press, 2006.
Comaroff, John L. "Symposium Introduction: Colonialism, Culture, and the Law: A Foreword." *Law & Social Inquiry* 26, no. 2 (2001): 305–14.
Commission, Jeffery. "The Duration and Costs of ICSID and UNCITRAL Investment Treaty Arbitrations." London: Vannin Capital, 2018. https://www.international-arbitration-attorney.com/wp-content/uploads/2018/07/Duration-and-Costs-of-ICSID-Arbitration.pdf.
Connell, Raewyn W. *The Men and the Boys*. St. Leonards, NSW: Allen & Unwin, 2000.
———. "The Sociology of Gender in Southern Perspective." *Current Sociology* 62, no. 4 (2014): 550–67.
———. "Why Is Classical Theory Classical?" *American Journal of Sociology* 102, no. 6 (1997): 1511–57.
Connell, Raewyn W., and James W. Messerschmidt. "Hegemonic Masculinity: Rethinking the Concept." *Gender & Society* 19, no. 6 (2005): 829–59.
Cooksey, Brian. "Politics, Patronage and Projects: The Political Economy of Agricultural Policy in Tanzania." Working Paper 40. Brighton: Future Agricultures Consortium, 2012.
———. "What Difference Has CAADP Made to Tanzanian Agriculture?" Working Paper 74. Brighton: Future Agricultures Consortium, 2013.
Cooper, Frederick. *Plantation Slavery on the East Coast of Africa*. Yale Historical Publications. Miscellany: 113. Yale University Press, 1977.
Coplan, David. "You Have Left Me Wandering About: Basotho Women and the Culture of Mobility." In *"Wicked" Women and the Reconfiguration of Gender in Africa*, edited by Dorothy Louise Hodgson and Sheryl McCurdy, 188–211. Portsmouth, NH: Heinemann, 2001.

Coquery-Vidrovich, Catherine. *African Women: A Modern History*. New York: Routledge, 1997.
Cornish, Gillian, and Rebekah Ramsay. "Gender and Livelihoods in Myanmar after Development-Induced Resettlement." *Forced Migration Review* 59 (2018): 55–57.
Cornwall, Andrea. "To Be a Man Is More Than a Day's Work: Shifting Ideals of Masculinity in Ado-Odo, Southwestern Nigeria." In *Men and Masculinities in Modern Africa*, edited by Lisa A. Lindsay and Stephan F. Miescher, 230–48. Social History of Africa. Portsmouth, NH: Heinemann, 2003.
Cotula, Lorenzo. "Land Grab or Development Opportunity? Agricultural Investment and International Land Deals in Africa." London: IIED, 2009.
———. "Land Rights and Investment Treaties: Exploring the Interface." London: IIED, 2015.
Cotula, Lorenzo, and Mika Schröder. "Community Perspectives in Investor-State Arbitration." Land, Investment and Rights Series. London: International Institute for Environment and Development (IIED), 2017.
Crenshaw, Kimberlé W. "Mapping the Margins: Intersectionality, Identity Politics, and Violence against Women of Color." *Stanford Law Review* 43, no. 6 (1991): 1241–99.
Cross, Charlotte. "Community Policing and the Politics of Local Development in Tanzania." *Journal of Modern African Studies* 52, no. 4 (December 2014): 517–40.
Dahir, Abdi Latif. "As Tanzania Votes, Many See Democracy Itself on the Ballot." *New York Times*, October 29, 2020, sec. World. https://www.nytimes.com/2020/10/28/world/africa/tanzania-election-john-magufuli.html.
Dancer, Helen. "An Equal Right to Inherit? Women's Land Rights, Customary Law and Constitutional Reform in Tanzania." *Social & Legal Studies* 26, no. 3 (2017): 291–310.
———. *Women, Land and Justice in Tanzania*. Woodbridge, UK: James Currey, 2015.
Dar24 Media. "Breaking News: Kamati Ya Katiba Na Sheria Bungeni Yakutana Kufanya Marekebisho Ya Sheria, Madini." 2017. https://www.youtube.com/watch?v=KQCNcyyrn2E.
Deere, Carmen Diana, and Magdalena León. "The Gender Asset Gap: Land in Latin America." *World Development* 31, no. 6 (2003): 925–47.
———. "Who Owns the Land? Gender and Land-Titling Programmes in Latin America." *Journal of Agrarian Change* 1, no. 3 (2001): 440–67.
Defelice, Michael S. "Yellow Nutsedge *Cyperus esculentus* L.: Snack Food of the Gods." *Weed Technology* 16, no. 4 (2002): 901–7.
Deininger, Klaus. "Land Policies for Growth and Poverty Reduction." Washington, DC: World Bank, 2003.
Deininger, Klaus, Derek Byerlee, Jonathan Lindsay, Andrew Norton, Harris Selod, and Mercedes Stickler. "Rising Global Interest in Farmland: Can It Yield Sustainable and Equitable Benefits?" Washington, DC: World Bank, 2011.
De Schutter, Olivier. "How Not to Think of Land-Grabbing: Three Critiques of Large-Scale Investments in Farmland." *Journal of Peasant Studies* 38, no. 2 (2011): 249–79.
De Soto, Hernando. *The Mystery of Capital: Why Capitalism Triumphs in the West and Fails Everywhere Else*. New York: Basic Books, 2000.
Development Today. "Filed Complaint about Sekab's Africa Venture." September 20, 2010.

———. "Internal Report: Scathing Criticism of Sida Support to Bagamoyo Bioenergy Project." May 2, 2017.

———. "Local Judge Says Municipality's Support for Sekab Was 'Illegal.'" January 5, 2012.

———. "Sekab 'Substantially Altered' Biofuel Study, Kept Orgut's Name." April 1, 2009.

Dlamini, Jacob. *Safari Nation: A Social History of the Kruger National Park*. Athens: Ohio University Press, 2020.

Dondeyne, Stephane, Els Vanthournout, John A. R. Wembah-Rashid, and Jozef Deckers. "Changing Land Tenure Regimes in Matrilineal Village of South Eastern Tanzania." *Journal of Social Development in Africa* 18, no. 1 (2003): 7–32.

Eberhardt, Pia, and Cecilia Olivet. "Profiting from Injustice: How Law Firms, Arbitrators and Financiers Are Fuelling an Investment Arbitration Boom." Brussels/Amsterdam: Corporate European Observatory and the Transnational Institute, 2012.

Edelman, Marc. "Bringing the Moral Economy Back in . . . to the Study of 21st-Century Transnational Peasant Movements." *American Anthropologist* 107, no. 3 (2005): 331–45.

Elgström, Ole. "Giving Aid on the Recipient's Terms: The Swedish Experience in Tanzania." In *Agencies in Foreign Aid: Comparing China, Sweden and the United States in Tanzania*, edited by Goran Hyden and Rwekaza Mukandala, 116–55. London: Palgrave Macmillan, 1999.

Englert, Birgit. "From a Gender Perspective: Notions of Land Tenure Security in the Uluguru Mountains, Tanzania." *Austrian Journal of Development Studies* 19, no. 1 (2003): 75–90.

Engström, Linda, and Flora Hajdu. "Conjuring 'Win-World'—Resilient Development Narratives in a Large-Scale Agro-investment in Tanzania." *Journal of Development Studies* 55, no. 6 (June 3, 2019): 1201–20.

Ergas, Zaki. "Why Did the Ujamaa Village Policy Fail?—towards a Global Analysis." *Journal of Modern African Studies* 18, no. 3 (1980): 387–410.

Escobar, Arturo. *Encountering Development: The Making and the Unmaking of the Third World*. Princeton, NJ: Princeton University Press, 1995.

EU. Directive 2003/30/EC of the European Parliament and of the Council of 8 May 2003 on the Promotion of the Use of Biofuels or Other Renewable Fuels for Transport, L 123 § (2003).

———. Directive 2009/28/EC of the European Parliament and of the Council of 23 April 2009 on the Promotion of the Use of Energy from Renewable Sources and Amending and Subsequently Repealing Directives 2001/77/EC and 2003/30/EC (2009).

Fabian, Steven. "East Africa's Gorée: Slave Trade and Slave Tourism in Bagamoyo, Tanzania." *Canadian Journal of African Studies / Revue canadienne des études africaines* 47, no. 1 (2013): 95–114.

———. *Making Identity on the Swahili Coast: Urban Life, Community, and Belonging in Bagamoyo*. African Identities: Past and Present. Cambridge: Cambridge University Press, 2019.

Fahim, Hussein M. *Dams, People, and Development: The Aswan High Dam Case*. Pergamon Policy Studies on International Development. New York: Pergamon, 1981.

Fairbairn, Madeleine. *Fields of Gold: Financing the Global Land Rush*. Ithaca, NY: Cornell University Press, 2020.

Faria, Caroline, and Sharlene Mollett. "Critical Feminist Reflexivity and the Politics of Whiteness in the 'Field.'" *Gender, Place & Culture* 23, no. 1 (2014): 79–93.

Federici, Silvia. *Caliban and the Witch: Women, the Body, and Primitive Accumulation.* New York: Autonomedia, 2004.

Feierman, Steven. *Peasant Intellectuals: Anthropology and History in Tanzania.* Madison: University of Wisconsin Press, 1990.

Feldman, Shelley, and Charles Geisler. "Land Expropriation and Displacement in Bangladesh." *Journal of Peasant Studies* 39, no. 3–4 (2012): 971–93.

Feldman, Shelley, Charles Geisler, and Louise Silbering. "Moving Targets: Displacement, Impoverishment, and Development." *International Social Science Journal* 175 (2003): 7–13.

Fent, Ashley, and Erik Kojola. "Political Ecologies of Time and Temporality in Resource Extraction." *Journal of Political Ecology* 27, no. 1 (January 21, 2020).

Ferguson, James. *The Anti-politics Machine: "Development," Depoliticization, and Bureaucratic Power in Lesotho.* Cambridge: Cambridge University Press, 1990.

———. *Expectations of Modernity : Myths and Meanings of Urban Life on the Zambian Copperbelt.* Berkeley: University of California Press, 1999.

———. *Give a Man a Fish: Reflections on the New Politics of Distribution.* Lewis Henry Morgan Lectures. Durham, NC: Duke University Press, 2015.

Foresight Group. "Göran Carstedt." 2020. https://www.foresight.se/gran-carstedt.

Forsyth, Tim. *Critical Political Ecology: The Politics of Environmental Science.* London: Routledge, 2003.

Foucault, Michel. *The History of Sexuality.* Vol. 1, *An Introduction.* New York: Pantheon Books, 1978.

———. *Society Must Be Defended: Lectures at the Collège de France, 1975–76.* New York: Picador, 2003.

Freidberg, Susanne. *French Beans and Food Scares: Culture and Commerce in an Anxious Age.* Oxford: Oxford University Press, 2004.

Freyhold, Michaela von. *Ujamaa Villages in Tanzania: Analysis of a Social Experiment.* New York: Monthly Review, 1979.

Friedmann, Harriet. "Patriarchy and Property: A Reply to Goodman and Redclift." *Sociologia Ruralis* 26, no. 2 (1986): 186–93.

Friends of the Earth Europe. "Ugandan Oil Palm Conglomerate Taken to Court over Land-Grab Claims." February 19, 2015. https://www.foei.org/press_releases/archive-by-subject/food-sovereignty-press/ugandan-oil-palm-conglomerate-taken-court-land-grab-claims.

Gagné, Marie. "Resistance against Land Grabs in Senegal: Factors of Success and Partial Failure of an Emergent Social Movement." In *The Politics of Land*, Research in Political Sociology vol. 26, edited by Tim Bartley, 173–203. Bingley, UK: Emerald, 2019.

Gallagher, Ed. "The Gallagher Review of the Indirect Effects of Biofuels Production." East Sussex, UK: Renewable Fuels Agency, 2008.

Gardner, Benjamin. *Selling the Serengeti: The Cultural Politics of Safari Tourism.* Athens: University of Georgia Press, 2016.

Geertz, Clifford. *The Interpretation of Cultures.* New York: Basic Books, 1973.

Geisler, Charles, and Fouad Makki. "People, Power, and Land: New Enclosures on a Global Scale." *Rural Sociology* 79, no. 1 (2014): 28–33.

Geschiere, Peter, and Stephen Jackson. "Autochthony and the Crisis of Citizenship: Democratization, Decentralization, and the Politics of Belonging." *African Studies Review* 49, no. 2 (2006): 1–7.

Geschiere, Peter, and Francis B. Nyamnjoh. "Capitalism and Autochthony: The Seesaw of Mobility and Belonging." *Public Culture* 12, no. 2 (2000): 423–52.

Gibson-Graham, J. K. *A Postcapitalist Politics*. Minneapolis: University of Minnesota Press, 2006.

Gilbert, Jérémie. "Strategic Litigation Impacts: Indigenous Peoples' Land Rights." New York: Open Society Foundation, 2018.

Glassman, Jonathon. *Feasts and Riot: Revelry, Rebellion, and Popular Consciousness on the Swahili Coast, 1856–1888*. Portsmouth, NH: Heinemann, 1995.

Global Arbitration Review. "Tanzania Faces ICSID Claim after Sugar Project Goes Sour." September 13, 2017.

GLOWS-FIU. "Water Atlas of the Wami/Ruvu Basin, Tanzania." North Miami: Global Water for Sustainability Program, Florida International University, 2014.

Gluckman, Max. "Gossip and Scandal." *Current Anthropology* 4, no. 3 (1963): 307–16.

Gonçalves, Euclides. "Orientações Superiores: Time and Bureaucratic Authority in Mozambique." *African Affairs* 112, no. 449 (October 1, 2013): 602–22.

Gorman, Timothy. "Moral Economy and the Upper Peasant: The Dynamics of Land Privatization in the Mekong Delta." *Journal of Agrarian Change* 14, no. 4 (2014): 501–21.

GRAIN. "Failed Farmland Deals: A Growing Legacy of Disaster and Pain." Against the Grain. Barcelona: GRAIN, 2018.

Grajales, Jacobo. "Land Grabbing, Legal Contention and Institutional Change in Colombia." *Journal of Peasant Studies* 42, no. 3–4 (July 4, 2015): 541–60.

Gramsci, Antonio. *Selections from the Prison Notebooks of Antonio Gramsci*. London: Lawrence and Wishart, 1999.

Green, Maia. "Trading on Inequality: Gender and the Drinks Trade in Southern Tanzania." *Africa* 69, no. 3 (1999): 404–25.

GTZ. *Liquid Biofuels for Transportation in Tanzania: Potential and Implications for Sustainable Agriculture and Energy in the 21st Century*. Berlin: German Technical Cooperation (GTZ), 2005.

Guardian. "CAG: Some Mining in Saadani National Park." April 19, 2022. https://www.ippmedia.com/en/news/cag-some-mining-saadani-national-park.

———. "Mkulazi Sugar Project to End Country's Sugar Deficit—Minister." December 18, 2018. https://www.ippmedia.com/en/news/mkulazi-sugar-project-end-country%E2%80%99s-sugar-deficit-minister.

———. "Police Sound Warning on Opposition Protests," November 2, 2020. https://www.ippmedia.com/en/news/police-sound-warning-opposition-protests%C2%A0.

———. "Revealed: Potentially 'Deadly' Fake Alcohol Being Sold across Tanzania." November 17, 2016. https://www.ippmedia.com/en/news/revealed-potentially-deadly-fake-alcohol-being-sold-across-tanzania.

Guthman, Julie. *Weighing In: Obesity, Food Justice, and the Limits of Capitalism*. Berkeley: University of California Press, 2011.

Guyer, Jane I. "Prophecy and the Near Future: Thoughts on Macroeconomic, Evangelical, and Punctuated Time." *American Ethnologist* 34, no. 3 (2007): 409–21.

Guyer, Jane I., and Pauline E. Peters. "Introduction: Conceptualizing the Household." *Development and Change* 18, no. 2 (1987): 197–214.

HabariLeo. "PM Assures Investors of Govt Protection." April 24, 2021. https://www.habarileo.co.tz/habari/2021-04-2760884f8b27fe4.aspx.

Hajer, Maarten A. *The Politics of Environmental Discourse: Ecological Modernization and the Policy Process*. Oxford: Oxford University Press, 1995.

Hall, Ruth, Marc Edelman, Saturnino M. Borras Jr., Ian Scoones, Ben White, and Wendy Wolford. "Resistance, Acquiescence or Incorporation? An Introduction to Land Grabbing and Political Reactions 'from Below.' Special Issue." *Journal of Peasant Studies* 42, no. 3–4 (2015): 467–88.
Halvorson-Quevedo, Raundi, and Mariana Mirabile. "Guarantees for Development." Paris: Organisation for Economic Co-operation and Development, 2014.
Hancock, Ange-Marie. *Solidarity Politics for Millennials: A Guide to Ending the Oppression Olympics*. The Politics of Intersectionality. New York: Palgrave Macmillan, 2011.
Harper, Douglas. "Talking about Pictures: A Case for Photo Elicitation." *Visual Studies* 17, no. 1 (2002): 13–26.
Hart, Gillian. "Development Critiques in the 1990s: Culs de Sac and Promising Paths." *Progress in Human Geography* 25, no. 4 (2001): 649–58.
———. "Engendering Everyday Resistance: Gender, Patronage and Production Politics in Rural Malaysia." *Journal of Peasant Studies* 19, no. 1 (October 1991): 93–121.
Harvey, David. *The Limits to Capital*. New and fully updated ed. London: Verso, 2006.
Havnevik, Kjell. *Tanzania: The Limits to Development from Above*. Uppsala: Nordiska Afrikainstitutet, 1993.
Havnevik, Kjell, and Hanne Haaland. "Biofuel, Land and Environmental Issues—the Case of Sekab's Biofuel Plans in Tanzania." In *Biofuels, Land Grabbing and Food Security in Africa*, edited by Prosper B. Matondi, Kjell Havnevik, and Atakilte Beyene, 106–33. London: Zed Books, 2011.
Havnevik, Kjell, Hanne Haaland, and Jumanne Abdallah. "Biofuel, Land, and Environmental Issues—the Case of Sekab's Biofuel Plans in Tanzania." Uppsala: Nordic Africa Institute, 2011.
Heiskanen, Veijo, and Laura Halonen. "Post-award Remedies." In *Litigating International Investment Disputes*, edited by Chiara Giorgetti, 497–526. Leiden: Brill, 2014.
Hines, Deborah A., and Karlyn Eckman. "Indigenous Multipurpose Trees of Tanzania: Uses and Economic Benefits for People." Rome: FAO, 1993.
Hittapunktse AB. "Carstedts Bil AB." 2021. https://www.hitta.se/carstedts+bil+ab/ume%C3%A5/kkmoovbij.
Hodgson, Dorothy Louise. *Being Maasai, Becoming Indigenous: Postcolonial Politics in a Neoliberal World*. Bloomington: Indiana University Press, 2011.
———. *Gender, Justice, and the Problem of Culture: From Customary Law to Human Rights in Tanzania*. Bloomington: Indiana University Press, 2017.
———. "'My Daughter . . . Belongs to the Government Now': Marriage, Maasai, and the Tanzanian State." *Canadian Journal of African Studies / Revue canadienne des études africaines* 30, no. 1 (1996): 106–23.
———. *Once Intrepid Warriors: Gender, Ethnicity, and the Cultural Politics of Maasai Development*. Bloomington: Indiana University Press, 2001.
Holt-Giménez, Eric. "The Biofuel Myth." *New York Times*, July 10, 2007.
hooks, bell. *Yearning: Race, Gender, and Cultural Politics*. New York: Routledge, 2015.
Hughes, Lotte. *Moving the Maasai: A Colonial Misadventure*. New York: Palgrave, 2006.
Human Rights Watch. "Tanzania: Freedoms Threatened Ahead of Elections." September 2, 2020. https://www.hrw.org/news/2020/09/02/tanzania-freedoms-threatened-ahead-elections.
Humphrey, Chris, and Annalisa Prizzon. "Guarantees for Development." London: Overseas Development Institute, 2014.

Hunsberger, Carol, and Alberto Alonso-Fradejas. "The Discursive Flexibility of 'Flex Crops': Comparing Oil Palm and Jatropha." *Journal of Peasant Studies* 43 (January 2, 2016): 225–50.

Hyden, Goran. *Beyond Ujamaa in Tanzania: Underdevelopment and an Uncaptured Peasantry*. Berkeley: University of California Press, 1980.

Ibbott, Ralph. *Ujamaa: The Hidden Story of Tanzania's Socialist Villages*. London: Crossroads Books, 2014.

ICSID. "Cases Database." 2020. https://icsid.worldbank.org/cases/case-database.

———. "EcoDevelopment in Europe AB & Others v. United Republic of Tanzania (ICSID Case No. ARB/17/33)." 2018. https://icsid.worldbank.org/cases/case-database/case-detail?CaseNo=ARB/17/33.

———. "The ICSID Caseload Statistics. Issue 2020-2." Washington, DC: International Center for Settlements of Investment Disputes, 2020. https://icsid.worldbank.org/resources/publications/icsid-caseload-statistics.

———. "Standard Chartered Bank (Hong Kong) Limited v United Republic of Tanzania (ICSID Case No. ARB/15/41)." Washington, DC: International Center for Settlements of Investment Disputes, 2019. https://icsid.worldbank.org/cases/case-database/case-detail?CaseNo=ARB/15/41.

IDI. "Case Brief: Class Action Lawsuit by Cambodian Villagers against Mitr Phol Sugar Corporation." 2018.

IFAD. *President's Report. Proposed Loan and Grant to the United Republic of Tanzania for the Bagamoyo Sugar Infrastructure and Sustainable Community Development Programme*. Executive Board, 115th Session, 15–16 September. Rome: IFAD, 2015.

IFC. "Performance Standard 5: Land Acquisition and Involuntary Resettlement." Washington, DC: IFC, 2012.

Iliffe, John. *A Modern History of Tanganyika*. Cambridge: Cambridge University Press, 1979.

IMF. "United Republic of Tanzania: Second Review under the Policy Support Instrument—Debt Sustainability Analysis." Washington, DC: International Monetary Fund, 2015.

Inbody, Megan. "The Political Role of the Gossip in Sweetnam the Woman-Hater, Arraigned by Women." In *The Politics of Female Alliance in Early Modern England*, edited by Christina Luckyj and Niamh J. O'Leary, 48–65. Lincoln: University of Nebraska Press, 2017.

Indra, Doreen Marie. *Engendering Forced Migration: Theory and Practice*. New York: Berghahn Books, 1999.

IRA. "Environmental Evaluation for Tourism Development in the Saadani Game Reserve." Dar es Salaam: Institute of Resource Assessment, University of Dar es Salaam, 1997.

Jacob, Thabit, and Rasmus Hundsbæk Pedersen. "New Resource Nationalism? Continuity and Change in Tanzania's Extractive Industries." *Extractive Industries and Society* 5, no. 2 (April 2018): 287–92.

James, R. W., and G. M. Fimbo. *Customary Land Law of Tanzania: A Source Book*. Nairobi: East African Literature Bureau, 1973.

Japhet, Kirilo, and Earle Seaton. *The Meru Land Case*. Nairobi: East African Publishing House, 1967.

Jenkins, Beth. "Mobilizing the Southern Agricultural Growth Corridor of Tanzania: A Case Study." Cambridge, MA: Corporate Social Responsibility Initiative, Harvard Kennedy School, 2012.

Jerman, Helena. *Between Five Lines: The Development of Ethnicity in Tanzania with Special Reference to the Western Bagamoyo District*. Uppsala: Finnish Anthropological Society and the Nordic Africa Institute, 1997.

Johansson, Emma Li, Marianela Fader, Jonathan W. Seaquist, and Kimberly A. Nicholas. "Green and Blue Water Demand from Large-Scale Land Acquisitions in Africa." *Proceedings of the National Academy of Sciences* 113, no. 41 (September 26, 2016): 11471–76.

Kabeer, Naila. *Reversed Realities: Gender Hierarchies in Development Thought*. London: Verso, 1994.

Kabendera, Erick, and Mark Anderson. "Tanzania Energy Scandal Ousts Senior Politicians." *Guardian*, December 24, 2014.

Kahoho, Timothy. "Ethanol Can Bring $30 Million Annually." JamiiForums, February 5, 2010. https://www.jamiiforums.com/threads/ethanol-can-bring-30-million-annually.52183/.

Kamanga, Khoti Chilomba. "The Agrofuel Industry in Tanzania: A Critical Enquiry into Challenge and Opportunities." Dar es Salaam: Land Rights Research and Resources Institute, 2008.

Kanyabwoya, Damas. "Magufuli's Method in 'Madness.'" *Citizen*, June 1, 2016. https://www.thecitizen.co.tz/tanzania/magazines/political-reforms/magufuli-s-method-in-madness--2556886.

Kapferer, Jean-Noël. *Rumors: Uses, Interpretations, and Images*. New Brunswick, NJ: Transaction, 1990.

Karashani, Bob. "Tanzania Reviews Law on Mediation for Investors." *East African*, February 8, 2020.

Kautsky, Karl. *The Agrarian Question*. London: Zwan, 1988.

Kdanka, Christopher. "Tanzania Ends Investment Treaty with Netherlands." *East African*, October 6, 2018. https://www.theeastafrican.co.ke/tea/business/tanzania-ends-investment-treaty-with-netherlands-1403968.

Kerkvliet, Benedict J., and James C. Scott, eds. *Everyday Forms of Peasant Resistance in South-East Asia*. New York: Routledge, 1986.

Kingdom of Sweden. "An Integrated Climate and Energy Policy," Prop.2008/09:162 and 163 (2009).

Kironde, J. M. Lusugga. "Improving Land Sector Governance in Africa: The Case of Tanzania." Washington, DC, 2009.

———. "The Regulatory Framework, Unplanned Development and Urban Poverty: Findings from Dar Es Salaam, Tanzania." *Land Use Policy* 23, no. 4 (2006): 460–72.

———. "Understanding Land Markets in African Urban Areas: The Case of Dar Es Salaam, Tanzania." *Habitat International* 24, no. 2000 (2000): 151–65.

Kitunga, Demere, and Marjorie Mbilinyi. "Rooting Transformative Feminist Struggles in Tanzania at Grassroots." *Review of African Political Economy* 36, no. 121 (2009): 433–41.

Koda, Bertha. "Changing Land Tenure Systems in the Contemporary Matrilineal Social System: The Gendered Dimension." In *The Making of a Periphery: Economic Development and Cultural Encounters in Southern Tanzania*, edited by P. Seppälä and B. Koda, 195–221. Uppsala: Nordiska Afrikainstitutet, 1998.

Kojima, Masami, and Todd Johnson. "Potential for Biofuels for Transport in Developing Countries." Washington, DC: World Bank, 2005.

Kombe, Wilbard J. "Land Acquisition for Public Use, Emerging Conflicts and Their Socio-Political Implications." *International Journal of Urban Sustainable Development* 2, no. 1–2 (2010): 45–63.

Krätli, Saverio, and Camilla Toulmin. *Farmer-Herder Conflict in Sub-Saharan Africa?* London: International Institute for Environment and Development, 2020.

Kusiluka, Moses Mpogole, Sophia Kongela, Moses Ayoub Kusiluka, Estron D. Karimuribo, and Lughano J. M. Kusiluka. "The Negative Impact of Land Acquisition on Indigenous Communities' Livelihood and Environment in Tanzania." *Habitat International* 35, no. 2011 (2011): 66–73.

Lal, Priya. *African Socialism in Postcolonial Tanzania: Between the Village and the World*. Cambridge: Cambridge University Press, 2015.

———. "Maoism in Tanzania: Material Connections and Shared Imaginaries." In *Mao's Little Red Book: A Global History*, edited by Alexander Cook, 96–116. Cambridge: Cambridge University Press, 2014.

Land Matrix. "Dynamics Overview Charts." 2021. https://landmatrix.org/charts/dynamics-overview/.

Lane, Charles. *Pastures Lost: Barabaig Economy, Resource Tenure, and the Alienation of Their Land in Tanzania*. London: IIED, 1994.

Lastarria-Cornhiel, Susana. "Impact of Privatization on Gender and Property Rights in Africa." *World Development* 25, no. 8 (1997): 1317–33.

Lay, Jann, Ward Anseeuw, Sandra Eckert, Insa Flachsbarth, Christoph Kubitza, Kerstin Nolte, and Markus Giger. "Taking Stock of the Global Land Rush: Few Development Benefits, Many Human and Environmental Risks. Analytical Report III." Centre for Development and Environment, University of Bern; Centre de coopération internationale en recherche agronomique pour le développement; German Institute of Global and Area Studies; University of Pretoria; Bern Open Publishing, September 23, 2021.

Lenin, Vladimir I. *The Development of Capitalism in Russia*. Moscow: Progress, 1967.

Li, Tania Murray. "To Make Live or Let Die? Rural Dispossession and the Protection of Surplus Populations." *Antipode* 41, no. S1 (2009): 66–93.

———. "What Is Land? Assembling a Resource for Global Investment." *Transactions of the Institute of British Geographers* 39, no. 4 (2014): 589–602.

———. *The Will to Improve: Governmentality, Development, and the Practice of Politics*. Durham, NC: Duke University Press, 2007.

Lindsay, Lisa A., and Stephan F. Miescher, eds. *Men and Masculinities in Modern Africa*. Social History of Africa. Portsmouth, NH: Heinemann, 2003.

Lindström, Jan. *Muted Memories: Heritage-Making, Bagamoyo, and the East African Caravan Trade*. New York: Berghahn Books, 2018.

Lissu, Tundu AM. "Tanzania's Five Years of Devastation under the Presidency of John Pombe Magufuli." *Daily Maverick*, March 22, 2021. https://www.dailymaverick.co.za/article/2021-03-22-tanzanias-five-years-of-devastation-under-the-presidency-of-john-pombe-magufuli/.

Little, Peter, and Michael Watts. *Living under Contract: Contract Farming and Agrarian Transformation in Sub-Saharan Africa*. Madison: University of Wisconsin Press, 1994.

Locher, Martina, and Ulrike Müller-Böker. "'Investors Are Good, If They Follow the Rules'—Power Relations and Local Perceptions in the Case of Two European Forestry Companies in Tanzania." *Geographica Helvetica* 69, no. 4 (2014): 249–58.

Locher, Martina, and Emmanuel Sulle. "Challenges and Methodological Flaws in Reporting the Global Land Rush: Observations from Tanzania." *Journal of Peasant Studies* 41, no. 4 (2014): 569–92.

Logan, Carolyn. "Ambitious SDG Goal Confronts Challenging Realities: Access to Justice Is Still Elusive for Many Africans." Afrobarometer Policy Paper No. 39, 2017.
Lugones, María. "Towards a Decolonial Feminism." *Hypatia* 25, no. 4 (2010): 742–59.
Lugongo, Bernard. "Bakhresa Kicks off Sugar Production." *Daily News*, December 31, 2017.
Machira, Polycarp. "50 Ditch CCM in Bagamoyo over Bitter Row with Investor." *Citizen*, May 15, 2013.
Maganga, Faustin, Kelly Askew, Rie Odgaard, and Howard Stein. "Dispossession through Formalization: Tanzania and the G8 Land Agenda in Africa." *Asian Journal of African Studies* 40 (2016): 3–49.
Mahmood, Saba. "Feminist Theory, Embodiment, and the Docile Agent: Some Reflections on the Egyptian Islamic Revival." *Cultural Anthropology* 16, no. 2 (2001): 202–36.
Makhulu, Anne-Maria. *Making Freedom: Apartheid, Squatter Politics, and the Struggle for Home*. Durham, NC: Duke University Press, 2015.
Makoye, Kizito. "Tanzania Seizes 'Idle' Land from Investors to Return to Poor Farmers." *Reuters*, June 10, 2016. https://www.reuters.com/article/us-tanzania-landrights-farming-idUSKCN0YN5GW.
———. "Villagers Spared Eviction as Tanzania Halts $500 Million Energy Project to Save Wildlife." *Reuters*, June 6, 2016. https://www.reuters.com/article/cnews-us-tanzania-investment-wildlife-idCAKCN0YS20T.
Malkki, Liisa H. "National Geographic: The Rooting of Peoples and the Territorialization of National Identity among Scholars and Refugees." *Cultural Anthropology* 7, no. 1 (1992): 24–44.
———. *Purity and Exile: Violence, Memory, and National Cosmology among Hutu Refugees in Tanzania*. Chicago: University of Chicago Press, 1995.
Mama, Amina A. "Sheroes and Villains: Conceptualising Colonial and Contemporary Violence against Women in Africa." In *Feminist Genealogies, Colonial Legacies, Democratic Futures*, edited by M. Jacqui Alexander and Chandra Talpade Mohanty, 46–62. New York: Routledge, 1997.
Marx, Karl. "The Eighteenth Brumaire of Louis Napoleon." In *The Marx-Engels Reader*, edited by Robert C. Tucker, 594–617. New York: W. W. Norton, 1978.
Masamba, Magalie. "Government Regulatory Space in the Shadow of BITs: The Case of Tanzania's Natural Resource Regulatory Reform." *Investment Treaty News* (blog), December 21, 2017. https://www.iisd.org/itn/2017/12/21/governmentregulatory-space-in-the-shadow-of-bits-the-case-of-tanzanias-natural-resource-regulatory-reform-magalie-masamba/.
Masquelier, Adeline. *Fada: Boredom and Belonging in Niger*. Chicago: University of Chicago Press, 2019.
Mawdsley, Emma, and Jack Taggart. "Rethinking d/Development." *Progress in Human Geography* 46, no. 1 (2022): 3–20.
Mbashiru, Katare. "Bakhresa Lands Sugar Covenant." *Tanzania Daily News*, October 7, 2016.
Mbembe, Achile. "Necropolitics." *Public Culture* 15, no. 1 (2003): 11–40.
Mbilinyi, Marjorie. "Analysing the History of Agrarian Struggles in Tanzania from a Feminist Perspective." *Review of African Political Economy* 43, no. sup1 (2016): 115–29.
———. "'City' and 'Countryside' in Colonial Tanganyika." *Economic and Political Weekly* 20, no. 43 (1985): WS88–96.

———. "Sweet and Sour: Women Working for Wages on Tanzania's Sugar Estates." In *How Africa Works: Occupational Change, Identity and Morality*, edited by Deborah Fahy Bryceson, 165–86. Rugby, UK: Practical Action, 2010.

McAuslan, Patrick. *Land Law Reform in Eastern Africa: Traditional or Transformative? A Critical Review of 50 Years of Land Law Reform in Eastern Africa 1961–2011.* Abingdon, UK: Routledge, 2013.

McCarthy, James. "A Socioecological Fix to Capitalist Crisis and Climate Change? The Possibilities and Limits of Renewable Energy." *Environment and Planning A: Economy and Space* 47, no. 12 (December 1, 2015): 2485–502.

McClintock, Anne. *Imperial Leather: Race, Gender, and Sexuality in the Colonial Contest.* New York: Routledge, 1995.

McDowell, Christopher, ed. *Understanding Impoverishment: The Consequences of Development-Induced Displacement.* New York: Berghahn Books, 1996.

McHenry, Dean E. *Tanzania's Ujamaa Villages: The Implementation of a Rural Development Strategy.* Berkeley: University of California Press, 1979.

McKay, Ben, Sérgio Sauer, Ben Richardson, and Roman Herre. "The Political Economy of Sugarcane Flexing: Initial Insights from Brazil, Southern Africa and Cambodia." *Journal of Peasant Studies* 43, no. 1 (January 2, 2016): 195–223.

McMichael, Philip. *Development and Social Change: A Global Perspective.* 7th ed. Thousand Oaks, CA: Sage, 2017.

Mehta, Lyla. "The Double Bind: A Gender Analysis of Forced Displacement and Resettlement." In *Displaced by Development: Confronting Marginalisation and Gender Injustice*, edited by Lyla Mehta, 3–33. New Delhi: Sage, 2009.

Meinzen-Dick, Ruth, and Esther Mwangi. "Cutting the Web of Interests: Pitfalls of Formalizing Property Rights." *Land Use Policy* 26, no. 1 (January 2009): 36–43.

Merchant, Carolyn. *The Death of Nature: Women, Ecology, and the Scientific Revolution.* New York: Harper & Row, 1980.

Midnight Notes Collective. "Introduction to the New Enclosures." *Midnight Notes* 10 (1990): 1–9.

Mies, Maria. *Patriarchy and Accumulation on a World Scale: Women in the International Division of Labour.* London: Zed Books, 1986.

Mignolo, Walter D. "DELINKING: The Rhetoric of Modernity, the Logic of Coloniality and the Grammar of De-coloniality." *Cultural Studies* 21, no. 2–3 (March 2007): 449–514.

Mitchell, Timothy. "Economics: Economists and the Economy in the Twentieth Century." In *The Politics of Method in the Human Sciences: Positivism and Its Epistemological Others*, edited by George Steinmetz, 126–41. Durham, NC: Duke University Press, 2005.

———. "Everyday Metaphors of Power." *Theory and Society* 19, no. 5 (1990): 545–77.

———. "The Limits of the State: Beyond Statist Approaches and Their Critics." *American Political Science Review* 85, no. 1 (1991): 77–96.

Mitchell, W. J. T. *Landscape and Power.* Chicago: University of Chicago Press, 2002.

Mohammed, Omar. "Twitter Is Having Fun with the Penny-Pinching Ways of Tanzania's New President." Quartz, November 26, 2015. https://qz.com/africa/560515/twitter-is-having-fun-with-the-penny-pinching-ways-of-tanzanias-new-president/.

Mollett, Sharlene, and Caroline Faria. "Messing with Gender in Feminist Political Ecology." *Geoforum* 45, no. 2013 (2013): 116–25.

Montgomery, Max. "Colonial Legacy of Gender Inequality: Christian Missionaries in German East Africa." *Politics & Society* 45, no. 2 (2017): 225–68.

Moore, Donald S. *Suffering for Territory: Race, Place, and Power in Zimbabwe*. Durham, NC: Duke University Press, 2005.

Mosse, David. *Cultivating Development: An Ethnography of Aid Policy and Practice*. London: Pluto, 2005.

Moyo, Sam, Paris Yeros, and Praveen Jha. "Imperialism and Primitive Accumulation: Notes on the New Scramble for Africa." *Agrarian South: Journal of Political Economy* 1, no. 2 (2012): 181–203.

Muchira, Njirani. "Tanzania Cites Bias as It Changes Laws Governing Arbitration." *East African*, September 17, 2018. https://www.theeastafrican.co.ke/tea/business/tanzania-cites-bias-as-it-changes-laws-governing-arbitration-1402606.

Mwagonde, Henry. "Bakhresa Firm to Start Sugar Production at Bagamoyo in 2 Years." *Guardian*, December 30, 2017. https://www.ippmedia.com/en/news/bakhresa-firm-start-sugar-production-bagamoyo-2-years.

Mwilongo, Patrick. "Prof Tibaijuka Aamuru Jengo La CCM Livunjwe." *Jambo Leo*, February 19, 2013.

Nachilongo, Hellen. "Water Scarcity Blow to Bakhresa Phase II Sugar Making Expansion Plans." *Citizen*, October 20, 2022. https://www.thecitizen.co.tz/tanzania/news/national/-water-scarcity-blow-to-bakhresa-phase-ii-sugar-making-expansion-plans-3991942.

Nagar, Richa. "The South Asian Diaspora in Tanzania: A History Retold." *Comparative Studies of South Asia, Africa and the Middle East* 16, no. 2 (1996): 62–80.

Namkwahe, John. "Tanzania's National Debt Stands at Sh55.43 Trillion, Govt Says It Is Sustainable." *Citizen*, June 11, 2020. https://www.thecitizen.co.tz/tanzania/news/national/tanzania-s-national-debt-stands-at-sh55-43-trillion-govt-says-it-is-sustainable-2710678.

Nash, Roderick. *Wilderness and the American Mind*. New Haven, CT: Yale University Press, 2014.

Neumann, Roderick P. *Imposing Wilderness: Struggles over Livelihood and Nature Preservation in Africa*. Berkeley: University of California Press, 1998.

Neville, Kate J. "The Contentious Political Economy of Biofuels." *Global Environmental Politics* 15, no. 1 (2015): 21–40.

New Alliance. "New Alliance for Food Security and Nutrition: 2013 Progress Report Summary." G8, 2013.

Ngcukaitobi, Tembeka. *The Land Is Ours: Black Lawyers and the Birth of Constitutionalism in South Africa*. Cape Town and Johannesburg: Penguin Random House South Africa, 2018.

Ng'hily, Dickson. "Air Tanzania Plane Seized in the Netherlands." *Citizen*, December 1, 2022. https://www.thecitizen.co.tz/tanzania/news/national/air-tanzania-plane-seized-in-the-netherlands-4039522.

Ng'wanakilala, Fumbuka. "Tanzania Says Resumes Talks for $10 Billion China-Backed Port." *Bloomberg*, June 26, 2021. https://www.bloomberg.com/news/articles/2021-06-26/tanzania-says-resumes-talks-for-10-billion-china-backed-port.

Nixon, Rob. *Slow Violence and the Environmentalism of the Poor*. Cambridge, MA: Harvard University Press, 2011.

Norris, Alison Holt, and Eric Worby. "The Sexual Economy of a Sugar Plantation: Privatization and Social Welfare in Northern Tanzania." *American Ethnologist* 39, no. 2 (2012): 354–70.

Nsehe, Mfonobong. "Tanzanian Tycoon Said Salim Bakhresa to Invest $30 Million in Zimbabwean Flour Mill." *Forbes*, June 29, 2016. https://www.forbes.com/sites/

mfonobongnsehe/2016/06/29/tanzanian-tycoon-said-bakhresa-to-invest-30-million-in-zimbabwean-flour-mill/.

Nyerere, Julius. "Ujamaa—the Basis of African Socialism." In *Ujamaa: Essays on Socialism*, 1–12. Oxford: Oxford University Press, 1968.

Oakland Institute. "The Looming Threat of Eviction: The Continued Displacement of the Maasai under the Guise of Conservation in Ngorongoro Conservation Area." Oakland, CA: Oakland Institute, 2021.

———. "Losing the Serengeti: The Maasai Land That Was to Run Forever." Oakland, CA: Oakland Institute, 2018.

———. "Understanding Land Deals in Africa: Tanzanian Villagers Pay for Sun Biofuels Investment Disaster." Land Deal Brief. Oakland, CA: Oakland Institute, 2012.

———. "Understanding Land Investment Deals in Africa. Country Report: Tanzania." Oakland, CA: Oakland Institute, 2011.

OECD. "Dispute Settlement Provisions in International Investment Agreements: A Large Sample Survey." Paris: OECD Investment Division, Directorate for Financial and Enterprise Affairs, 2012.

———. "OECD Investment Policy Reviews: Tanzania 2013." Paris: Organisation for Economic Co-operation and Development, 2013.

Ogden, Laura A. *Swamplife: People, Gators, and Mangroves Entangled in the Everglades*. Minneapolis: University of Minnesota Press, 2011.

Ogola, George. "#Whatwouldmagufulido? Kenya's Digital 'Practices' and 'Individuation' as a (Non)Political Act." *Journal of Eastern African Studies* 13, no. 1 (January 2, 2019): 124–39.

OHCHR. "In Her Global Human Rights Update, Bachelet Calls for Urgent Action to Heighten Resilience and Protect People's Rights." September 14, 2020. https://www.ohchr.org/EN/NewsEvents/Pages/DisplayNews.aspx?NewsID=26226.

Omar, Hazla. "NSSF, PPF Link Up to Establish Sugar Factory." *Tanzania Daily News*, October 21, 2016. https://allafrica.com/stories/201610210519.html.

Orozco-Quintero, Alejandra, and Leslie King. "A Cartography of Dispossession: Assessing Spatial Reorganization in State-Led Conservation in Saadani, Tanzania." *Journal of Political Ecology* 25, no. 2018 (2018): 40–63.

Ortner, Sherry B. "Resistance and the Problem of Ethnographic Refusal." *Comparative Studies in Society and History* 37, no. 1 (1995): 173–93.

Our World in Data. "Goal 5: Gender Equality—SDG Tracker." SDG Tracker. https://sdg-tracker.org/gender-equality.

Owere, Paul. "President Samia Reshuffles Cabinet, Drops Chief Secretary." *Citizen*, April 1, 2021. https://www.thecitizen.co.tz/tanzania/news/president-samia-reshuffles-cabinet-drops-chief-secretary-3343728.

Oxfam Tanzania. "Balancing Infrastructure Development and Community Livelihoods: Lessons from Mtwara–Dar es Salaam Natural Gas Pipeline." Dar es Salaam: Oxfam Tanzania, 2017.

Paget, Dan. "Again, Making Tanzania Great: Magufuli's Restorationist Developmental Nationalism." *Democratization* 27, no. 7 (October 2, 2020): 1240–60.

Pailey, Robtel Neajai. "De-centring the 'White Gaze' of Development." *Development and Change* 51, no. 3 (2019): 729–45.

Pedersen, Rasmus Hundsbæk, and Opportuna Kweka. "The Political Economy of Petroleum Investments and Land Acquisition Standards in Africa: The Case of Tanzania." *Resources Policy* 52 (2017): 217–25.

Peluso, Nancy L. *Rich Forests, Poor People: Resource Control and Resistance in Java*. Berkeley: University of California Press, 1992.

Peters, Pauline E. "'Our Daughters Inherit Our Land, but Our Sons Use Their Wives' Fields': Matrilineal-Matrilocal Land Tenure and the New Land Policy in Malawi." *Journal of Eastern African Studies* 4, no. 1 (2010): 179–99.
Peterson, Luke E. "Land Deals Could Sow Arbitration Disputes." *Kluwer Arbitration Blog*, May 30, 2009. http://arbitrationblog.kluwerarbitration.com/2009/05/30/land-deals-could-sow-arbitration-disputes/.
Phillips, Kristin D. *An Ethnography of Hunger: Politics, Subsistence, and the Unpredictable Grace of the Sun.* Bloomington: Indiana University Press, 2018.
Pietilä, Tuulikki. *Gossip, Markets, and Gender: How Dialogue Constructs Moral Value in Post-socialist Kilimanjaro.* Madison: University of Wisconsin Press, 2007.
Pitcher, M. Anne. "Forgetting from Above and Memory from Below: Strategies of Legitimation and Struggle in Postsocialist Mozambique." *Africa* 76, no. 1 (2006): 88–112.
Plumwood, Val. *Feminism and the Mastery of Nature.* London: Routledge, 1993.
Polanyi, Karl. *The Great Transformation: The Political and Economic Origins of Our Time.* 3rd ed. Boston: Beacon, 2001.
Policy Forum. "Tanzania Governance Review 2014: The Year of 'Escrow.'" Dar es Salaam: Policy Forum, 2015.
———. "Tanzania Governance Review 2015–16: From Kikwete to Magufuli: Break with the Past or More of the Same?" Dar es Salaam: Policy Forum, 2016.
Popkin, Samuel. *The Rational Peasant: The Political Economy of Rural Society in Vietnam.* Berkeley: University of California Press, 1979.
Quijano, Anibal. "Coloniality of Power and Eurocentrism in Latin America." *Nepantla: Views from the South* 1, no. 3 (2000): 533–80.
Rabobank. "Tanzania Sugar. Rabobank Industry Note #386." Utrecht: Rabobank International, 2013.
Radcliffe, Sarah A. *Dilemmas of Difference: Indigenous Women and the Limits of Postcolonial Development Policy.* Durham, NC: Duke University Press, 2015.
———. "Gendered Frontiers of Land Control: Indigenous Territory, Women and Contests over Land in Ecuador." *Gender, Place & Culture* 21, no. 7 (2013): 854–71.
Ranger, Terence. "The Invention of Tradition in Colonial Africa." In *The Invention of Tradition*, edited by E. J. Hobsbawm and Terence Ranger, 211–62. Cambridge: Cambridge University Press, 1983.
Reuters. "Tanzania, Brazil Sign Biofuel Memo, Talk Debt Relief." July 7, 2010.
Ribot, Jesse C., and Nancy Peluso. "A Theory of Access." *Rural Sociology* 68, no. 2 (2003): 153–81.
Rist, Gilbert. *The History of Development: From Western Origins to Global Faith.* London: Zed Books, 2008.
Roberntz, P., T. Edman, and A. Carlson. "The Rufiji Landscape: The Sweet and Bitter Taste of Sugarcane Grown for Biofuel." Stockholm: WWF Sweden, 2009.
Rockel, Stephen J. *Carriers of Culture: Labor on the Road in Nineteenth-Century East Africa.* Social History of Africa Series. Portsmouth, NH: Heinemann, 2006.
Rodney, Walter. *How Europe Underdeveloped Africa.* London: Verso Books, 2018.
Roseberry, William. "Hegemony and the Language of Contention." In *Everyday Forms of State Formation: Revolution and the Negotiation of Rule in Modern Mexico*, edited by G. Joseph and D. Nugent, 355–65. Durham, NC: Duke University Press, 1996.
Rwegasira, Abdon. *Land as a Human Right: A History of Land Law and Practice in Tanzania.* Dar es Salaam: Mkuki na Nyota, 2012.

Rweyemamu, Aisia. "The Cost of JPM's Economic Warfare." *Guardian*, May 28, 2017. https://www.ippmedia.com/en/news/cost-jpm%E2%80%99s-economic-warfare.

Rysman, Alexander. "How the 'Gossip' Became a Woman." *Journal of Communication* 27, no. 1 (1977): 176–80.

SAGCOT. "SAGCOT Investment Blueprint." Dar es Salaam: SAGCOT, 2011.

———. "SAGCOT Investment Partnership Program. Opportunities for Investors in the Sugar Sector." Dar es Salaam: Southern Agricultural Corridor of Tanzania, 2012.

Schipani, Andres. "Tanzania's New President Turns Her Back on Magufuli." *Financial Times*, April 14, 2021. https://www.ft.com/content/7549086b-515f-44ea-850c-9795d4a124d6.

Schippers, Mimi. "Recovering the Feminine Other: Masculinity, Femininity, and Gender Hegemony." *Theory and Society* 36, no. 1 (March 9, 2007): 85–102.

Schlimmer, Sina. "Caught in the Web of Bureaucracy? How 'Failed' Land Deals Shape the State in Tanzania." *International Development Policy* 12, no. 1 (January 1, 2020).

Schönweger, Oliver, and Peter Messerli. "Land Acquisition, Investment, and Development in the Lao Coffee Sector: Successes and Failures." *Critical Asian Studies* 47, no. 1 (2015): 94–122.

Schroeder, Rick. *Africa after Apartheid: South Africa, Race, and Nation in Tanzania*. Bloomington: Indiana University Press, 2012.

———. *Shady Practices: Agroforestry and Gender Politics in the Gambia*. Berkeley: University of California Press, 1999.

Scoones, Ian, Ruth Hall, Saturnino M. Borras Jr., Ben White, and Wendy Wolford. "The Politics of Evidence: Methodologies for Understanding the Global Land Rush. Special Issue." *Journal of Peasant Studies* 40, no. 3 (2013): 469–83.

Scott, James C. *Domination and the Arts of Resistance: Hidden Transcripts*. New Haven, CT: Yale University Press, 1990.

———. *Seeing Like a State: How Certain Schemes to Improve the Human Condition Have Failed*. New Haven, CT: Yale University Press, 1998.

———. *The Moral Economy of the Peasant: Rebellion and Subsistence in Southeast Asia*. New Haven, CT: Yale University Press, 1976.

———. *Weapons of the Weak: Everyday Forms of Peasant Resistance*. New Haven, CT: Yale University Press, 1985.

Scott, Joan W. "The Evidence of Experience." *Critical Inquiry* 17, no. 4 (1991): 773–97.

Scotton, Carol M. "Some Swahili Political Words." *Journal of Modern African Studies* 3, no. 4 (December 1965): 527–41.

Sekab. "Ethanol as a Raw Material." n.d. https://www.sekab.com/en/ethanol-as-a-raw-material/.

———. "Our History." Sekab, 2020. https://www.sekab.com/en/about-us/about-the-company/our-history/.

Sekab BT. "Application for Credit Enhancement Guarantee. A Letter to the Swedish International Development Agency, Dated 28 July. Ref: ABE/0900728/01," 2009.

———. "Environmental and Social Impact Statement of the Proposed BioEthanol Production on the Former Razaba Ranch, Bagamoyo District, Tanzania. Final Report. Submitted to National Environmental Management Council." Dar es Salaam: Sekab BioEnergy Tanzania, 2008.

Shaidi, Leonard P. "Crime, Justice and Politics in Contemporary Tanzania: State Power in an Underdeveloped Social Formation | Office of Justice Programs." *International Journal of the Sociology of Law* 17, no. 3 (1989): 247–71.

Sheriff, Abdul. *Slaves, Spices and Ivory in Zanzibar: Integration of an East African Commercial Empire into the World Economy, 1770–1873*. Oxford: Oxford University Press, 1987.
Shibutani, Tamotsu. *Improvised News: A Sociological Study of Rumor*. Indianapolis: Bobbs-Merrill, 1966.
Shiva, Vandana. *Staying Alive: Women, Ecology, and Survival in India*. New Delhi: Women Unlimited, 1988.
Shivji, Issa G. "Lawyers in Neoliberalism: Authority's Professional Supplicants or Society's Amateurish Conscience?" *CODESRIA Bulletin*, no. 3 and 4 (2006): 15–25.
———. *Not Yet Democracy: Reforming Land Tenure in Tanzania*. London: International Institute for Environment and Development, 1998.
———. *State Coercion and Freedom in Tanzania*. Human and People's Rights Monograph Series, No. 8. Roma: Institute of Southern African Studies, National University of Lesotho, 1990.
Shome, Raka. "Whiteness and the Politics of Location: Postcolonial Reflections." In *Whiteness: The Communication of Social Identity*, edited by Judith Martin and Thomas K. Nakayama, 107–28. Thousand Oaks, CA: Sage, 1999.
Sida. "Decision on Contribution to EcoEnergy Tanzania." Stockholm: Swedish International Development Agency, February 28, 2014.
———. "Letter RE: Sekab BioEnergy Tanzania Ltd—Application for Credit Enhancement Guarantee, Reference Number: 2009–001283." 2009.
———. "Sida Avslutar Stödet till EcoEnergy." 2015.
Sitari, Taimi. "Settlement Changes in the Bagamoyo District of Tanzania as a Consequence of Villagization." *Fennia* 161, no. 1 (1983): 1–90.
Smith, Dorothy. "Women, Class and Family." *Socialist Register* 20 (1983): 1–44.
Snyder, Katherine A. "Mothers on the March: Iraqi Women Negotiating the Public Sphere in Tanzania." *Africa Today* 53, no. 1 (2006): 79–99.
Songa, Songa wa. "Bagamoyo Embarks on Model Sugar Production Programme." *Citizen*, March 17, 2014.
———. "Key Financiers Drop Sh1tr Sugar Project." *Citizen*, April 25, 2015.
———. "Sh1 Trillion Sugar Deal Vanishes into Thin Air." *Citizen*, April 25, 2015.
Spivak, Gayatri Chakravorty. "Explanation and Culture: Marginalia." In *The Spivak Reader: Selected Works of Gayatri Chakravorty Spivak*, edited by Donna Landry and Gerald MacLean, 29–52. New York: Routledge, 1996.
———. "Subaltern Studies: Deconstructing Historiography." In *The Spivak Reader: Selected Works of Gayatri Chakravorty Spivak*, edited by Donna Landry and Gerald MacLean, 203–36. New York: Routledge, 1996.
Stasik, Michael, Valerie Hänsch, and Daniel Mains. "Temporalities of Waiting in Africa." *Critical African Studies* 12, no. 1 (January 2, 2020): 1–9.
State House. "Rais Dkt John Magufuli Amtembelea Mama Maria Nyerere Nyumbai Kwake Msaani." 2018. https://www.youtube.com/watch?v=jqZeOgY7tNk.
Stein, Howard, Faustin P. Maganga, Rie Odgaard, Kelly Askew, and Sam Cunningham. "The Formal Divide: Customary Rights and the Allocation of Credit to Agriculture in Tanzania." *Journal of Development Studies*, 2016: 1–14.
Stoler, Ann. *Capitalism and Confrontation in Sumatra's Plantation Belt, 1870–1979*. Ann Arbor: University of Michigan Press, 1995.
———. "Imperial Debris: Reflections on Ruins and Ruination." *Cultural Anthropology* 23, no. 2 (2008): 191–219.
———. *Race and the Education of Desire: Foucault's History of Sexuality and the Colonial Order of Things*. Durham, NC: Duke University Press, 1995.

Stoller, E. W., and R. D. Sweet. "Biology and Life Cycle of Purple and Yellow Nutsedges (*Cyperus rotundus* and *C. esculentus*)." *Weed Technology* 1, no. 1 (1987): 66–73.

Stoller, E. W., and L. M. Wax. "Yellow Nutsedge Shoot Emergence and Tuber Longevity." *Weed Science* 21, no. 1 (1973): 76–81.

Sulle, Emmanuel. "Bureaucrats, Investors and Smallholders: Contesting Land Rights and Agro-commercialisation in the Southern Agricultural Growth Corridor of Tanzania." *Journal of Eastern African Studies* 14, no. 2 (April 2, 2020): 332–53.

———. "Social Differentiation and the Politics of Land: Sugar Cane Outgrowing in Kilombero, Tanzania." *Journal of Southern African Studies*, 2016: 1–17.

Sulle, Emmanuel, and Helen Dancer. "Gender, Politics and Sugarcane Commercialisation in Tanzania." *Journal of Peasant Studies* 47, no. 5 (July 28, 2020): 973–92.

Sulle, Emmanuel, and Fred Nelson. "Biofuels, Land Access and Rural Livelihoods in Tanzania." London: International Institute for Environment and Development, 2009.

Sunseri, Thaddeus. "The Baumwollfrage: Cotton Colonialism in German East Africa." *Central European History* 34, no. 1 (March 2001): 31–51.

———. "Famine and Wild Pigs: Gender Struggles and the Outbreak of the Majimaji War in Uzaramo (Tanzania)." *Journal of African History* 38, no. 2 (July 1997): 235–59.

Swantz, Marja-Liisa. *Blood, Milk, and Death: Body Symbols and Power of Regeneration among the Zaramo of Tanzania*. Westport, CT: Bergin & Garvey, 1995.

———. *Ritual and Symbol in Transitional Zaramo Society with Special Reference to Women*. Lund, Sweden: C W K Fleerup, 1970.

TANAPA. "Saadani National Park Management Zone Plan Environmental Impact Assessment." Arusha: Tanzania National Parks Authority, 2003.

Tanganyika Territory. Land Ordinance, 1923.

Taussig, Michael T. *The Devil and Commodity Fetishism in South America*. Chapel Hill: University of North Carolina Press, 1980.

TAWLA. "Position Paper on Gender Mainstreaming of the Constitution Review Process of Tanzania." Dar es Salaam: Tanzania Women Lawyers Association, 2013.

Tenga, Ringo W. "Legitimizing Dispossession: The Tanzanian High Court's Decision on the Eviction of Maasai Pastoralists from Mkomazi Game Reserve." *Cultural Survival Quarterly* 22, no. 4 (Winter 1998).

Tenga, Ringo W., and George Kakoti. "The Barabaig Land Case: Mechanics of State-Organised Land-Grabbing in Tanzania." *Centre for Development Research / International Work Group for Indigenous Affairs Document* 74, no. 1993 (1993): 39–51.

Tenga, Ringo W., and Sist J. Mramba. "Manual on Land Law and Conveyancing in Tanzania." Dar es Salaam: Tumaini University, 2008.

Thompson, Katrina Daly. "The Embodied Socialization of a Muslim Bride in Zanzibar Town." In *Gendered Lives in the Western Indian Ocean: Islam, Marriage, and Sexuality on the Swahili Coast*, edited by Erin E. Stiles and Katrina Daly Thompson, 168–208. Athens: Ohio University Press, 2015.

TNI. "Licensed to Grab: How International Investment Rules Undermine Agrarian Justice." Briefing. The Hague: Transnational Institute, 2015.

———. "A Transatlantic Corporate Bill of Rights." The Hague: Transnational Institute, 2013.

Trudeau, Daniel. "Politics of Belonging in the Construction of Landscapes: Place-Making, Boundary-Drawing and Exclusion." *Cultural Geographies* 2006, no. 13 (2006): 421–43.

Tsikata, Dzodzi. "Securing Women's Interests within Land Tenure Reforms: Recent Debates in Tanzania." *Journal of Agrarian Change* 3, no. 1–2 (2003): 149–83.

Tsing, Anna Lowenhaupt. *Friction: An Ethnography of Global Connection*. Princeton, NJ: Princeton University Press, 2005.

———. *The Mushroom at the End of the World: On the Possibility of Life in Capitalist Ruins*. Princeton, NJ: Princeton University Press, 2015.

———. "Supply Chains and the Human Condition." *Rethinking Marxism* 21, no. 2148–176 (2009): 30.

Tuck, Eve. "Suspending Damage: A Letter to Communities." *Harvard Educational Review* 79, no. 3 (September 1, 2009): 409–28.

Turner, Victor. "Variations on a Theme of Liminality." In *Secular Ritual*, edited by Sally Falk Moore and Barbara G. Myerhoff, 36–52. Assen, Netherlands: Van Gorcum, 1977.

UNCTAD. "Fact Sheet on Investor-State Dispute Settlement Cases in 2018." IIA Issue Note. Geneva: United Nations Conference on Trade and Development, May 2019. https://unctad.org/system/files/official-document/diaepcbinf2019d4_en.pdf.

———. "Investment Policy Hub. Investment Dispute Settlement Navigator," 2020. https://investmentpolicy.unctad.org/investment-dispute-settlement/country/222/united-republic-of-tanzania.

———. "Nachingwea and Others v. Tanzania," 2020. https://investmentpolicy.unctad.org/investment-dispute-settlement/cases/1086/nachingwea-and-others-v-tanzania.

URT (United Republic of Tanzania). *Annual General Report of the Controller and Auditor General on the Audit of Public Authorities and Other Bodies for the Financial Year 2020/21*. 2022.

———. Budget Digest 2004/5. 2004.

———. Constitution of the United Republic of Tanzania. 1977.

———. Declaration of Saadani National Park. Government Notice No. 281. 2005.

———. Hotuba ya Wizara ya Ardhi, Nyumba, na Maendeleo ya Makazi 2012/13. 2012.

———. Kilimo Kwanza Resolution. 2009.

———. Land Acquisition Act. 1967.

———. Land Act. 1999.

———. Land (Assessment of the Value of Land for Compensation) Regulations. 2001.

———. Land (Compensation Claims) Regulations. 2001.

———. People's Militia (Powers of Arrest) Act. 1975.

———. Regional Administration Act. 1997.

———. Speech by the Minister of Finance and Planning, Hon. Dr. Philip I. Mpango (MP), Presenting to the National Assembly the Estimates of Government Revenue and Expenditure for 2020/21. 2020.

———. Tanzania Development Vision 2025. 1999.

———. Tanzania Investment Act. 1997.

———. Village Land Act. 1999.

———. "Waziri Mkuu: Watanzania Tubadilike, Tujivunie Vyetu. Akagua Mradi Wa Kiwanda Cha Sukari Na Shamba La Miwa Bagamoyo." Press Release, August 18, 2020. https://www.pmo.go.tz/index.php/news/pmnews/567-waziri-mkuu-watanzania-tubadilike-tujivunie-vyetu.

———. Wildlife Conservation (Game Reserves) Order, 1974 (G.Ns. Nos. 265 and 275 of 1974).

US Department of State. "2020 Investment Climate Statements: Tanzania." Washington, DC: US Department of State, 2020. https://www.state.gov/reports/2021-investment-climate-statements/tanzania/.
———. "2021 Investment Climate Statements: Tanzania." Washington, DC: US Department of State, 2021. https://www.state.gov/reports/2021-investment-climate-statements/tanzania/.
Velde, Dirk Willem te, and Isabella Massa. "Donor Responses to the Global Financial Crisis—a Stock Take." Global Financial Crisis Discussion Series Paper 11. London: Overseas Development Institute, 2009.
Vhugen, Darryl, Jennifer Duncan, Laura Eshbach, and Rugemeleza Nshala. "Approaches to Agricultural Investment Models, Valuation and Support for Local Communities in Tanzania." Dar es Salaam: USAID SERA Policy Project, 2014.
Vuorela, Ulla. *The Women's Question and the Modes of Human Production: An Analysis of a Tanzanian Village.* Uppsala: Nordiska Afrikainstitutet, 1987.
Walker, Peter, and L. Fortmann. "Whose Landscape? A Political Ecology of the 'Exurban' Sierra." *Cultural Geographies* 2003, no. 10 (2003): 469–91.
Walsh-Dilley, Marygold. "Negotiating Hybridity in Highland Bolivia: Indigenous Moral Economy and the Expanding Market for Quinoa." *Journal of Peasant Studies* 40, no. 4 (2013): 659–82.
Walwa, William John. "Land Use Plans in Tanzania: Repertoires of Domination or Solutions to Rising Farmer–Herder Conflicts?" *Journal of Eastern African Studies* 11, no. 3 (July 3, 2017): 408–24.
Wang, Caroline, and Mary Ann Burris. "Photovoice: Concept, Methodology, and Use for Participatory Needs Assessment." *Health Education & Behavior* 24, no. 3 (1997): 369–87.
Watts, M. *Silent Violence: Food, Famine, and Peasantry in Northern Nigeria.* Berkeley: University of California Press, 1983.
WEF. "Grow Africa: Partnering to Achieve African Agriculture Transformation." Geneva: World Economic Forum in collaboration with A. T. Kearney, 2016.
———. "Realizing a New Vision for Agriculture: A Roadmap for Stakeholders." Geneva: World Economic Forum in collaboration with McKinsey & Company, 2010.
Weil, Simone. *The Need for Roots: Prelude to a Declaration of Duties towards Mankind.* London: Routledge, 2005.
White, Ben, Saturnino M. Borras Jr., Ruth Hall, Ian Scoones, and Wendy Wolford. "The New Enclosures: Critical Perspectives on Corporate Land Deals. Special Issue." *Journal of Peasant Studies* 39, no. 3–4 (2012): 619–47.
White, Luise. *Speaking with Vampires: Rumor and History in Colonial Africa.* Berkeley: University of California Press, 2001.
Whitehead, A. "I'm Hungry Mum: The Politics of Domestic Budgeting." In *Of Marriage and the Market*, edited by Kate Young, Carol Wolkowitz, and Roslyn McCullagh, 88–111. London: CSE Books, 1981.
Whitehead, A., and Dzodzi Tsikata. "Policy Discourses on Women's Land Rights in Sub-Saharan Africa: The Implications of the Re-turn to the Customary." *Journal of Agrarian Change* 2, no. 1–2 (2003): 67–112.
Wilen, C. A., M. E. McGiffen, and C. L. Elmore. "Nutsedge." Pest Notes. Davis: Agriculture and Natural Resources Program, University of California, 2010.
Williams, Raymond. *Marxism and Literature.* Oxford: Oxford University Press, 1977.
Wilson, Anika. *Folklore, Gender, and AIDS in Malawi: No Secret under the Sun.* Gender and Cultural Studies in Africa and the Diaspora. New York: Palgrave Macmillan, 2013.

Wise, Timothy A. "Picking up the Pieces from a Failed Land Grab Project in Tanzania." *GlobalPost* (blog), June 27, 2014. https://www.pri.org/stories/2014-06-27/picking-pieces-failed-land-grab-project-tanzania.

———. "What Happened to the Biggest Land Grab in Africa? Searching for ProSavana in Mozambique." *Foodtank* (blog), 2014. https://foodtank.com/news/2014/12/what-happened-to-the-biggest-land-grab-in-africa-searching-for-prosavana-in/.

Wolford, Wendy. *This Land Is Ours Now: Social Mobilization and the Meanings of Land in Brazil*. Durham, NC: Duke University Press, 2010.

Wolford, Wendy, Saturnino M. Borras Jr., Ruth Hall, Ian Scoones, and Ben White. "Governing Global Land Deals: The Role of the State in the Rush for Land. Special Issue." *Development and Change* 44, no. 2 (2013): 189–210.

World Bank. "Tanzania: New World Bank Financing to Secure Land Rights for Up to Two Million Citizens." December 21, 2021. https://www.worldbank.org/en/news/press-release/2021/12/21/tanzania-new-world-bank-financing-to-secure-land-rights-for-up-to-two-million-citizens.

———. *World Development Report 2008: Agriculture for Development*. Washington, DC: World Bank, 2007.

WWF Tanzania. "Biofuel Industry Study, Tanzania. An Assessment of the Current Situation." Dar es Salaam: World Wildlife Fund for Nature–Tanzania Programme Office, 2008.

Yngstrom, Ingrid. "Women, Wives and Land Rights in Africa: Situating Gender beyond the Household in the Debate over Land Policy and Changing Tenure Systems." *Oxford Development Studies* 30, no. 1 (2002): 21–40.

Young, Iris Marion. "The Logic of Masculinist Protection: Reflections on the Current Security State." *Signs* 29, no. 1 (2003): 1–25.

Zheng, Sarah. "China, Tanzania in Talks to Get US$10 Billion Bagamoyo Port Project Back on Track, Ambassador Says." *South China Morning Post*, July 11, 2019. https://www.scmp.com/news/china/diplomacy/article/3018217/china-tanzania-talks-get-us10-billion-bagamoyo-port-project.

Index

Page numbers in *italics* refer to figures.

Abu-Lughod, Lila, 149
ActionAid, 30, 48, 150
African Development Bank (AfDB), 2, 41, 44–49, 89–90, 93–94, 104, 107
African National Congress, 150
African socialism. See *ujamaa*
African Union, 41, 43
Afro-Shirazi Party, 77, 192n74
agricultural development, 3, 5, 27, 40, 43, 79, 171, 175; commercialization, 17–18, 40–44. See also Bakhresa project; EcoEnergy Sugar Project; outgrower schemes; public-private partnership (PPP)
Agricultural Sector Development Program (ASDP), 41–43
Agro EcoEnergy Tanzania, 37
alcohol brewing, by women, 126, 137–40, *138*
Alliance for a Green Revolution in Africa, 42
Alpers, Edward, 199n15
Anzaldúa, Gloria, 12
arbitration. See international arbitration
Arusha Declaration (1967), 75, 173
Askew, Kelly, 175
authoritarianism, 171, 173–74

Bachelet, Michelle, 174
Bagamoyo District Council, 149
Bagamoyo EcoEnergy Limited, 37
Bagamoyo Sugar Limited. See Bakhresa project
Bagamoyo town, 1, 21, 53–54, 74, 80, 110, 142, 170, 180, 194n94, 194n97
Bakhresa, Said Salim, 169, 211n4
Bakhresa project, 169–82, 211nn3–5, 212n29, 213n45, 213n60
Barabaig (ethnic group), 83, 150, 189n8. See also pastoralists
Beban, Alice, 192n54
belonging, 6, 14, 17–18, 55, 94, 124, 128, 144–45. See also citizenship
Berger, John, 53
Bergfors, Anders, 36–37
Berry, Sara, 171

Big Results Now (BRN), 38, 42
BioAlcohol Fuel Foundation, 29, 195n15
biofuels (ethanol), 1, 3, 26–31, 36–39, 195n15, 197n66
biopolitical governmentality, 11, 106–12, 122. See also Early Measures
Bondei (ethnic group), 100
Borras, Saturnino M., Jr., 215n84
Bozi (settlement), 80–82, 98, 105, 110, 130, 138, 144, 146, 165, 180, 202n83
Brazil, 29, 38
Butler, Judith, 14

capitalism, 5–12, 17, 171, 183–84; "double movement" and, 51–52; feminist critique of, 14–16. See also development; global financial crisis (2008)
Carstedt, Göran, 28
Carstedt, Per, 28–30, 33–34, 36, 38–39, 45, 50–51, 88–89, 104, 154, 195n15, 197n58, 206n47
cattle ranching. See RAZABA ranch
CCM (Chama Cha Mapinduzi), 16, 77, 80, 123–24, 144, 173–74, 179–80, 192n74, 198n95; in Makaani, 149–50, 155, 158–60
census. See People and Property Count (PPC)
Cernea, Michael, 203n1
certificates of customary rights of occupancy (CCROs), 83, 99, 102, 153. See also land rights: customary
Chachage, Chambi, 193n81
Chadema, 144, 174, 179
Chagga, 161
charcoal making, 99, 115, 117–18, 121, 126, 132, 134–36, *137*, 138–39, 142–43
Christianity. See missionaries
citizenship, 14, 17, 128, 145; agrarian, 142–44, 147. See also belonging
Civic United Front, 80
civil and political rights, 141, 150, 167. See also disenfranchisement
civil society, 28, 198n119
clandestine farming, 126, 132–34, 208n32

243

INDEX

class, 5, 14–15, 20, 145, 189n6, 194nn95–96, 212n29. *See also* elites; identities; poverty
climate change mitigation, 1, 28–29, 36, 38–39. *See also* biofuels (ethanol)
colonialism, 9–10, 19, 197n73; British, 53, 73–75; culture of legality, 209n15; gender and, (*see* gender: coloniality of); German, 1, 53, 56, 71–74; legacies of, 54–56, *55*, *57–58*, 60, 69, 71–72, *72*, 75–76, 162–63, 191n50, 194n96, 201n45, 207n4; migrant labor and, 62, 162–63; taxation, 15, 74. *See also* enclosure; plantations
Comaroff, Jean and John, 209n15
communal farming. *See ujamaa*: communal farming
Community Finance Company (CFC), 29, 195n15, 195n19, 196n21
compensation, 2, 13; budget constraints, 103; of compassion, 93, 178; delayed, 92, 103–4, 124, 127; eligibility for, 4, 20, 93–94, 144–45, 176–79, 204n16; estimated value, 95, 204n17; gender inequality in payment (*see* gender: in valuation and compensation); international best practices for, 90, 94, 96, 104; by TANAPA, 84–86, 128, 166, 203n114; Tanzanian laws on, 32–33, 46, 93. *See also* land compensation fund; People and Property Count (PPC)
Comprehensive Africa Agriculture Development Programme, 41
compulsory land acquisition, 93, 96, 132, 176; national retirement schemes and, 211n3; public interest and, 13, 32, 92, 183. *See also* land: Tanzanian laws and policies on
conservation. *See* wildlife conservation
contentious politics, 19, 125, 150, 182. *See also* resistance
contract farming. *See* EcoEnergy Sugar Project: outgrower scheme; outgrower schemes
corruption, 11, 49–50, 77, 116–17, 141, 172
COVID-19 pandemic, 174–75
credit enhancement guarantee, 36–37, 47
criminality, 117, 132, 134. *See also* illicit practices
critical agrarian studies, 18–19, 193n87
crops, 61–62, *63*, 65; cash, 71, 74, 88, 138, 176; flexible ("flex"), 39; food (including fruit trees), 54, 60–62, *63–65*, 64–66, 72, 75, 81–82, 99, 133–34, 162, 178, 185; gender and, 62. *See also* land: use restrictions; subsistence

de Schutter, Olivier, 184
de Soto, Hernando, 103

development: coloniality of, 10; distinction between "big D" and "little d," 9–10, 191n42; paternalism and, 9, 40; privatization of, 36, 41, 48. *See also* agricultural development; capitalism; public-private partnership (PPP)
development agencies, 30, 36, 43, 211n5, 214n77. *See also* German Technical Cooperation Agency (GTZ); International Fund for Agricultural Development (IFAD); Swedish International Development Cooperation Agency (Sida); United States Agency for International Development (USAID)
disenfranchisement, 18, 165–67
displacement: development-induced, 96, 203nn1–2; ex situ, 5, 184; in situ, 6–10, 166, 184. *See also* dispossession; involuntary resettlement
dispossession: by Bakhresa project, 179–81; land formalization and, 91, 103; safety net against, 98, 102; by TANAPA, 84, 86–88, 154, 158–60, 166–68; uncertainty, 4–9, 51, 105, 124, 127–28, 147 (*see also* liminality). *See also* displacement; involuntary resettlement
distributive justice, 143–44
divorce, 99, 102, 103, 111, 156, 166. *See also* single mothers; women: "without men"
Doe (ethnic group), 57, 59, 61, 157

Early Measures, 106–8, *113*, 127; gender and, 109–13, 122. *See also* biopolitical governmentality; foreign development consultant; microfinance loans; Resettlement Action Plan
EcoDevelopment in Europe AB, 28, 36–37, 195n14, 197n58, 297n66
EcoEnergy Africa AB, 37
EcoEnergy Sugar Project (formerly Sekab BioEnergy Project): cancellation, 4, 12, 128, 169–71, 211n8; capital accumulation, 27; claims of sustainability, 30, 35; community knowledge about, 92, 105; comparison to TANAPA, 87–88; concession boundaries, *3*, 106, 113, 114, *115*, 116; concession size, 13, 20, 34, 46, 198n107; criticism of, 87–88, 101, 124, 134, 141, 145–47, 198n119 (*see also* resistance); delays and uncertainty, 4–5, 8, 11–12, 23, 46, 48–52, 81, 90, 103–5, 112, 121, 124–27, 140, 146, 166; employees, 34, 37, 115, 119–21, 135–36, 165, 170, 206n46; as "failure," 4, 11; financing, 36–37, 44–48, 90, 107, 112–13, 197n57, 211n5; governance

of, 105–22; labor conditions, 142–43; land acquisition process, 12–13, 33–35, 44–46, 52, 89; launch event, 1–5; map of, 2, 22, 70; outgrower scheme, 2–3, 9, 38, 40, 47–48, 130; ownership structure, 37; partnership with Tanzanian government, 13, 30, 34, 45, 48–52, 90, 189n3; population resettlement (*see* Resettlement Action Plan); socio-environmental impacts of, 30–31, 35–37, 88, 181–82, 189n7, 199n119 (*see also* environmental and social impact assessment (ESIA))
electricity, 2, 47, 197n66
elites, 17, 19, 38, 43, 49, 88, 156–57, 160–61, 165, 173, 179; among the poor (*see* Makaani: male elders)
enclosure: colonial-era, 71–75; cycles of, 5, 185; feminist analysis of, 14–15; "new," 10; partial or incomplete, 5, 10–12, 52; postcolonial, 75–88. *See also* land grabbing
Enemy Property (Disposal) Proclamation (1920), 73
energy sector: global, 28–29, 190n29; in Sweden, 28, 35–36, 195n3; in Tanzania, 2, 29–30, 47, 49, 51, 72, 175, 182, 212n23, 214n73. *See also* biofuels; electricity
environmental and social impact assessment (ESIA), 35, 44, 88, 189n7. *See also* Eco Energy Sugar Project: socio-environmental impacts of
ethnic groups in Tanzania, 15, 59, 61, 66, 157, 189n8, 200n21, 200n32. See also *specific ethnic groups*
ethnography, 18–23, 194n92, 205n32

familyhood. See *ujamaa*
female solidarity, 62, 68
femininity, 14, 106, 112. *See also* gender: norms; women: initiation rites
feminist political ecology, 14
Ferguson, James, 10
fishing, 13, 59, 64, 88, 181, 214n73
flood events, 76, 99, 158–59
floodplain agriculture, 13, 54, 59–65, 63–64, 73, 76–77, 152, 154, 164, 176, 180–81. See also *kitopeni;* Wami River
food production, household. See subsistence
food sovereignty, 183
foreign aid (overseas development assistance), 36, 41
foreign development consultant, 90–91, 96, 105–9, 113, 116, 122, 127–28, 206n26. *See also* Early Measures; Resettlement Action Plan

foreign investors, 1–3, 43, 46, 172, 175, 190n31, 212n18. *See also* EcoEnergy Sugar Project; international arbitration; public-private partnership (PPP)
Foucault, Michel, 18, 106
Franco, Jennifer C., 215n84
Friedmann, Harriet, 98
fuelwood collection, by women, 7, 65, 101, 110, 131, 139

G8, 43
Gama (settlement), 72, 75–76, 87; boundaries, 152. *See also* Makaani (settlement); trespass lawsuit
Geertz, Clifford, 194n92
gender: coloniality of, 14–16; defined, 14; domesticity and care work, 14–15, 17, 20, 95–98, 110, 138, 145 (*see also* labor: gender division of); in land access and control, 15, 61, 91, 95–96, 99, 111, 117–18, 151, 157–58, 163–64, 183, 200n21; norms, 15, 20, 106, 110–11, 117, 122, 138, 145, 214n65; in political participation, 21, 145–46; in valuation and compensation, 91, 94–103, 111–12, 131, 177–79. *See also* Early Measures: gender and; femininity; identities; masculinity; men; patriarchy; women
general land (category), 31–32, 35, 46
German Technical Cooperation Agency (GTZ), 29, 84–85
Gibson-Graham, J. K., 191n42
global financial crisis (2008), 36, 41
global land grab. *See* land grabbing
Gogo (ethnic group), 80, 98. *See also* RAZABA ranch
gongo, 134, 136–37, 139–40
gossip, 125–26, 129–32, 207n21
GRAIN, 11
Gramsci, Antonio, 18, 143, 208n57
Grow Africa, 43
guerrilla agriculture. *See* clandestine farming

Ha (ethnic group), 99
HakiArdhi, 195n15
Hall, Ruth, 150
Hancock, Ange-Marie, 14
Hart, Gillian, 9, 191n42
Hassan, Samia Suluhu, 174–75, 190n31
High Court of Tanzania, 80; patriarchal disposition of, 164. *See also* trespass lawsuit
Hodgson, Dorothy, 193n90
hooks, bell, 184, 207n10
human rights, 93, 151, 174, 176, 206n47

246   INDEX

humor, 130–31
Hut and Poll Tax Ordinance (1922), 74

identities: as coalitional, 147–48; as co-constituted with landscapes and livelihoods, 55–69; as complex and intersectional, 14, 19, 109, 114–15, 122, 142–48, 151, 167, 181; spatial, 17, 148; strategic essentialism, 18. *See also* belonging; class; gender; race; social positions
Iliffe, John, 73
illicit practices, 125–26, 132–40, *137*, 147
in-betweenness, 5–9. *See also* liminality
informants (local farmers): Abdallah, 141–42, 145; Amina, 98–99, 102, 103; Athumani, 6, 144; Aziza, 133–34, 137; Bambadi, 149–50, 152–61, *155*, 159–61, 165–67, 177, 210n17, 213n60; Beatrice, 119; Daudi, 119–22, 129–30, 185; Emmanuel, 131, 142–43, 146–47; Fatuma, 161–64, *163*, 166; Hadija, 157–58, 166; Halima, 62–64, *63*, 112; Hasani, *163*–64; Jennifa, 145–46; John, 142; Juma, 60–61; Majid, 141, 146–47; Mariamu, 139–40; Meena, 117; Mohammedi, 127–30, 178–79; Mosi, 105, 109; Musa, 133–34; Mwajuma, 60–61, 64, *64*–65, 86, *86*, 178–79, 181; Mwanahamisi, 179, 185; Neema, *7*, 62, 138–39, *138–39*; Nuru, *8*, 100–103; Omari, 130; Rama, 124, 141, 147, 177–80, 214n65; Rukia, 99–100, 102, 103, 135–36, 156; Samwel, 141–42, 145; Selemani, *118*, *155*; Shabani, 120–21, 140; Suba, 176, 177, 180; Tatu, 139–40; Thabiti, 152–55, 158–64, *163*, 165–67; Yusuf, 110–12, 146–47; Zainab, 110–12; Zuberi, 119
inheritance, 15, 61, 152, 158, 161–63, 166, 200n21
international arbitration, 12, 214nn72–73; EcoEnergy case, 170–71, 182; against Tanzania, 182–84; Tanzanian prohibitions on, 172, 212n23. *See also* International Centre for Settlement of Investment Disputes (ICSID)
International Centre for Settlement of Investment Disputes (ICSID), 170–71, 181, 183, 211nn14–15, 214n73
international financial institutions (IFIs), 41, 44, 90, 92, 94, 211n5. *See also* African Development Bank (AfDB); International Fund for Agricultural Development (IFAD); World Bank
International Fund for Agricultural Development (IFAD), 2, 41, 48

"intruders" or "squatters," 13, 18, 83, 93–94, 115, 132, 134, 143, 165–67, 176
investment treaties, 170–71, 182–84, 212n23. *See also* international arbitration
involuntary resettlement: defined, 90; international best practices on, 90, 94, 104, 106–7, 203n1. *See also* displacement; dispossession; Resettlement Action Plan
Islam, 15, *67*, 68, 111, 158, 200n21, 200n36. See also *maulidi*

Jumbe, Aboud, 77

Karume, Abeid, 77
Kawambwa, Shukuru, 123–24
kiangazi, 62, *63*
Kikwete, Jakaya, 3, 23, 34, 37, 41–43, 49, 80, 172, 174, 195n15, 212nn17–18; son of, 88
Kilimo Kwanza initiative, 41–43
Kilombero Sugar Company, 3, 189n6, 197n61
Kisauke (subvillage): displaced villagers, 154, 158–59, 161–64, 166–67; history of, *55*, 69, *72*, 72–75, 201n45; Saadani National Park and, 84–86, 158. *See also* Matipwili (village)
Kitame (subvillage), 56, 85, 87, 152, 158–59, 163
kitopeni, 62–64, *63*, 65, 200n23
Kitunga, Demere, 90
Kombe, Wilbard, 94
Kutu (ethnic group), 59
Kwere (ethnic group), 59, 66, 100

labor: colonial demands, 14, 15, 62–71, 162–63; EcoEnergy and (*see* EcoEnergy Sugar Project: labor conditions); gender division of, 61–62, 81, 110, 137; land grabbing and, 13, 29; missionaries and, 54; socialist-era demands (see *ujamaa*)
land: as commodity and property, 13, 89; "empty," "idle," or "unused," 12, 32–35, 55, 73, 82, 172; feminist ontology of, 192n54; land-water as coupled resource, 55, 62–65, 200n28; Tanzanian laws and policies on, 13, 31–35, 44, 46, 75, 83, 93, 97, 153, 191n50, 193n81; use restrictions, 7, 93, 106, 116, 132, 141
Land Acquisition Act (1967), 31, 93
Land Act (1999), 31–33, 46, 83, 93, 97
land bank, 32–34, 44
land compensation fund, 33, 44, 97
"land for equity," 1, 44–46, 189n3
Land Form 69a (government form), 95, 99–102, 177, 204n18. *See also* compensation; People and Property Count (PPC)

INDEX

land formalization, 90–104
land grabbing, 3, 10–11, 19, 51; activism against, 31, 35, 48 (*see also* resistance); cases in Tanzania, 11; international arbitration and, 182–84; land formalization and, 103; legitimating narratives, 27, 39–40, 44, 51, 172–73; litigation as response to, 150–51. *See also* enclosure
Land Matrix, 10, 190n29
Land Ordinance (1923), 73–74
land rights: customary, 15, 46, 61, 74–75, 83, 93, 116, 147, 153, 162, 202n90 (*see also* certificates of customary rights of occupancy (CCROs)); discrepancy between land access and, 12–13; exclusive (free of encumbrances), 1, 44–46, 88, 116, 122; planting of trees and, 6, 8, 60–61, 75, 81–82; struggles for (*see* resistance); techno-legal tools and, 89; violation of, 149–50; women's (*see* gender: in land access and control). *See also* "intruders" or "squatters"; land title; property
land sales, 100, 155–57, 160–61, 194n94
landscape: defined, 14; making of Bagamoyo's, 53–89; remaking of Bagamoyo's, 175–82
land sovereignty, 124, 143–44, 183
land survey, 12–13, 33–34, 71, 89, 150, 152, 159, 198n107
land title: customary (*see* certificates of customary rights of occupancy (CCROs)); EcoEnergy's project, 46, 89, 115–16, 170; gender bias in titling, 91; programs for (*see* land formalization)
lawfare, 151, 158, 164, 166–67, 209n15
Leipziger Baumwollspinnerei (LBS), 71–73, 83
liminality, 5–9, 11–12, 18, 181, 191n39, 191n42
Lissu, Tundu, 174
Luguru (ethnic group), 59, 66
Lukuvi, William, 172

Magufuli, John Pombe, 49, 141, 145, 146, 169, 179, 198n91, 212nn17–18; socialist nostalgia and, 171–76
Mahmood, Saba, 19
Maji Maji uprising, 71–72
Makaani (settlement): community development in, 124, 155, 160, 166; history of, *58*, 59, 61, 162; male elders, 91–92, 100, *118*, 149, *155*. *See also* Gama (settlement); trespass lawsuit
Makua (ethnic group), 59, 162

Makurunge (village): history of, 76; maps of, *78*; RAZABA ranch and, 81; township status, 56. *See also* Kitame (subvillage); Razaba (subvillage)
Maoism, 16, 75
Marxism, 9–10, 16, 190n14
masculinity, 14, 106, 108, 114–17, 214n65; masculization of space, 20, 117, 122. *See also* gender; men
masika, 61–62, *163–64*
Matipwili (village): access to, 87–88, 194n97; confrontation with TANAPA (*see* Saadani Game Reserve; Saadani National Park; Wildlife Division); farming in, 59, 63, *64*, 64–65 (*see also* floodplain agriculture); history of, 57–59, 76, 162–63; maps of, 56–57, *57*, 69–71, *70*, *78*
matriliny, 15, 61, 66, 157, 200n21, 200n32. *See also mkole*; women: initiation rites
maulidi, 67, 68, 200n36
Mauritius, 37, 197n57
Mbembe, Achille, 106, 114
Mbilinyi, Marjorie, 90
men: exclusionary politics and, 149–68; gossip and, 129; as husbands, 4, 15, 91, 95–101, 109–12, 117–18, 131, 133, 151, 157, 164, 166–67, 177, 179, 194n98; political organizing and, 140–41, 144 (see also *wanaharakati* (activist group)). *See also* gender; masculinity; patriarchy
mgambo: bribery and, 117, 136; establishment of, 113–14; photograph of, *118*; physical violence, 7, 10, 24–25, 105–6, 114, 117–24, 206n47; rumors and gossip about, 127–30; security checkpoints, 114, *115*, 138; surveillance, 122, 126, 132–34, 142–43; threats and intimidation tactics, 7, 106, 116–19, 122, 124, 126–27, 142, 177. *See also* necropolitical governmentality
microfinance loans, 107–9. *See also* Early Measures
mining, 172–73, 182–83, 190n29, 214n73. *See also* resource nationalism
Ministry of Agriculture, Food Security, and Cooperatives (Ministry of Agriculture), 3, 38, 43, 195n19
Ministry of Energy and Minerals, 30, 195n19
Ministry of Lands, Housing, and Human Settlements (Ministry of Lands), 32, 34, 45, 88–90, 92–93, 96, 132, 159, 172, 175, 195n19
Ministry of Natural Resources and Tourism, 84, 88, 89, 194n97

INDEX

Ministry of Planning, Economy, and Empowerment (Ministry of Planning), 30, 34, 195n19
missionaries, 15, 199n6, 200n21; Holy Ghost Mission in Bagamoyo, 53, 199n6
Mkapa, Benjamin, 41, 212n17
mkole, 65–68, *67–68*, 72, 95, 200n32. *See also* women: initiation rites
modernization, 3, 9–10, 27, 40, 42, 184
Mto wa Ngoma, 72, 85
Mvave River, 84

National Agricultural and Food Corporation, 79
National Audit Office, 181
National Biofuel Task Force, 30, 195n19
National Business Council, 43
National Electoral Commission, 174
National Environmental Management Council, 35
National Ranching Corporation, 79
National Service, 114
National Strategy for Growth and the Reduction of Poverty (MKUKUTA), 41, 197n77
necropolitical governmentality, 106, 114–22. *See also mgambo*
neoliberalism, 5, 9, 17, 41, 56, 83, 102, 173, 175, 182, 193n81
New Alliance for Food Security and Nutrition, 43
New Vision for Agriculture, 42
ngoma, 61, 68. *See also mkole*; women: initiation rites
Norwegian government, 30, 43
nutrition transition, 38–39
Nyamwezi (ethnic group), 53, 59, 98
Nyerere, Julius, 16, 75–77, 79, 84, 113, 145, 173, 176. *See also* Arusha Declaration (1967); *ujamaa*

Operesheni Pwani, 75–77
Operesheni Vijiji, 75
ordinary speech acts, 125–32, 147
Ortner, Sherry, 194n92
outgrower schemes, 17, 42, 51, 189n6. *See also* EcoEnergy Sugar Project: outgrower scheme

paramilitary forces. *See mgambo*
pastoralists, 4, 80, 83, 88, 98, 178, 180, 189n8, 214n66
patriarchy, 15, 74, 90, 98, 103, 129, 151, 167, 197n73; nuclear family and, 15, 17, 94–98, 102, 177, 200n21, 205n29

patriliny, 15, 61, 157, 200n21
peasant intellectuals, 143, 208n57
Peluso, Nancy, 13
People and Property Count (PPC), 90–104, 150; forms used in, 95, 99–102, 177, 204n18; rejection of, 91, 117, 156; reuse of data, 176
People's Militia Act (1975), 114
Phillips, Kristin, 17
photovoice, 21–23, 194n99, 195n100
plantations, 18, 40, 42, 51, 60, 162, 199n119; cotton, 56, 71–73, 83; sisal, 56, 73–75, 162, 202n101. *See also* agricultural development; outgrower schemes; sugarcane
Pogoro (ethnic group), 165
Polanyi, Karl, 18, 51, 214n81
political organizing, 123–26, 140–48
polygyny, 97, 102, 111–12
postcolonial nation building, 5, 9, 16, 56, 75, 96, 114, 128, 173, 176, 192n81; enclosures and, 75–88
postsocialism, 9, 17, 175
poverty, 8, 96, 102–3, 110, 115–17, 126, 132, 145–46, 156, 158, 160, 177, 181, 183, 203n1
Prime Minister's Office, 4, 31–32, 38, 43, 45, 49, 79–80, 141, 169–70, 177, 213n48
Project Affected Persons ("PAPs"), 93–94, 106–8, 122
property: assertion of, 106, 116; labor theory of, 61, 147, 162; valuation (*see* People and Property Count (PPC)). *See also* land; land formalization; land rights; land title
Puar, Jasbir, 105
public-private partnership (PPP), 2, 27–28, 40, 41–44, 48, 50–51
Puri, Arvind, 37

Quijano, Aníbal, 14

race, 5, 14, 19–20, 29, 40, 109, 194nn95–96, 212n29; colonialism and, 13, 19, 173, 207n4; ethnographic encounter and, 205n32; exclusion of minorities, 17, 173. *See also* whiteness
Razaba (subvillage), 56, 61, 82, 159; activist group in (*see wanaharakati*); elections, 141, 165; farming and livelihood activities in, 62, *63*; history of, 61, 77–83, 159; primary school, 81, *81–82*
RAZABA ranch, 33–35, 77–83, 86–87, 159–60, 201n71; boundaries, 69, *70*, 79; workers at, 80–81, 98
renewable energy, 28–29. *See also* biofuels
reserved land (category), 31

INDEX

Resettlement Action Plan, 4, 44, 90–92, 106–7, 189n8. *See also* Early Measures; foreign development consultant; involuntary resettlement
resistance, 18–19, 125, 185, 207n10; as adaptation, 125; everyday acts of presence, 5, 14, 52, 55–56, 82, 125, 185; formal-legal, 150–51; moral economies of, 130–32, 142, 144, 147; power and, 19, 193n86; repertoires of, 126–48 (*see also* illicit practices; lawfare; ordinary speech acts; political organizing); social difference and, 122, 142, 144–48, 193n90; state responses to, 142–43, 149, 160, 177
resource nationalism, 171–73, 175
Ribot, Jesse, 13
Rufiji District, 33–34
rumor, 125–28, 169–70
Ruvu River, *22*, 54, *70*

Saadani (village), 53, *78*, 84, 202n94
Saadani Game Reserve, 83–85, 202n101. *See also* Saadani National Park; Tanzania National Parks Authority (TANAPA); Wildlife Division
Saadani National Park, 69, *70*, 83–89, *86*, 87, 154, 158, 166–69, 181–82, 194n97. *See also* Saadani Game Reserve; Tanzania National Parks Authority (TANAPA); Wildlife Division
Scott, James, 125, 126
Sekab (Svensk Etanol Kemi AB), 28–31
Sekab BT (Sekab BioEnergy Tanzania), 1, 30–31, 33–37, 196n43, 197n66. *See also* EcoEnergy Sugar Project
Shivji, Issa, 114
Sinare, Amani, 195n15
single mothers, 95, 99, 103, 120, 126, 145, 156, 165–66. *See also* divorce; widows; women: "without men"
slave trade, 14–15, 53–54, 199n4
smallholder farmers, 16–19, 88, 158, 160, 193n89; as obstacle to progress, 16, 17; persistence of, 5, 14, 52, 55–56, 185; relationship to capitalist agriculture, 3, 9–10, 17, 38, 43 (*see also* agricultural development; outgrower schemes). *See also* floodplain agriculture; resistance; subaltern agency
Smith, Dorothy, 11
social difference. *See* identities; resistance: social difference and; social positions
social positions, 122, 125, 151, 193n90; as compared to positionings, 18, 193n90

social reproduction, 13, 56, 65, 89; crisis of, 71–72, 114–15, 181, 225; gender and, 81, 106, 109–12, 115, 122 (*see also* gender: domesticity and care work)
Southern Agriculture Corridor of Tanzania (SAGCOT), 42–43
spatiotemporality, 5, 8–9, 191n39. *See also* liminality
Standard Bank of South Africa (SBSA), 47–48
State House, 34, 43, 50
Stoler, Ann, 54
subaltern agency, 16–19, 151, 193n89. *See also* identities; resistance; social positions
subsistence, 17, 51, 62, 72, 76, 81, 101, 128, 185, 208n32, 247. *See also* crops
sugar: imports, 49, 153–54, 213n60; import substitution of, 38–39, 47, 153–54, 169, 213n60; national policy on, 49
sugarcane: donor interests in, 30; as flexible commodity, 39; investor interests in, 1–3, 27, 33, 39–40, 169, 197n61, 197n66, 211n3; Tanzanian government interest in, 38, 42. *See also* Bakhresa project; biofuels; EcoEnergy Sugar Project
Swantz, Marja-Liisa, 200n32
Sweden: ambassador to Tanzania, 2–3; relationship with Tanzania, 2, 182
Swedish International Development Cooperation Agency (Sida), 2, 30, 36–37, 46–50

Tanganyika African National Union (TANU), 77, 192n74; Youth League, 113
Tanzania Investment Center (TIC), 32–34, 46, 196n36
Tanzanian Agricultural Development Bank, 211n5
Tanzania National Parks Authority (TANAPA), 83, 85–88, 128, 166–67, 194n97, 202n92. *See also* Saadani Game Reserve; Saadani National Park; Wildlife Division
Tanzanian government: criticism of, 30, 50–51, 71, 76, 85, 124, 134, 170, 174; EcoEnergy and (*see* EcoEnergy Sugar Project: partnership with Tanzanian government; "land for equity"); landownership, 13, 18, 31, 75, 83, 191n50, 193n81; national budget, 214n77; national debt, 41, 82, 85, 183, 214n77; parliament, 4, 44–45, 88, 173, 174
Taussig, Michael, 18
taxes, 37–38, 49, 62, 132, 172. *See also* colonialism: taxation

250 INDEX

Tengwe (subvillage), 75, 84
Tibaijuka, Anna, 44–45, 49, 149, 198n95
tourism sector, 84–88, 175
trees: multiple uses of, 12–13, 65 (*see also* charcoal making; crops; fuelwood collection, by women; *mkole*); photographs of, *6, 8, 63, 86*; restrictions on cutting down, 115–16, 135 (see also *mgambo*); women's cultural traditions and (see *mkole*)
trespass lawsuit, 149–68
Tsing, Anna, 27, 191n44, 197n73
Tuck, Eve, 184
Turner, Victor, 11–12

ujamaa, 9, 16–17, 75, 173, 175–76, 198n81; communal farming, 16, 71, 75–77, 101; villagization, 16–17, 56, 75–77, *78*, 83, 113–14. See also Arusha Declaration (1967); Nyerere, Julius
United Nations, 38, 174, 184, 198n95, 212n18; Private Sector Forum, 43. *See also* International Fund for Agricultural Development (IFAD)
United States Agency for International Development (USAID), 45–46
Uttam Group, 48

valuation. *See* People and Property Count (PPC)
Valuation Form 1 (government form), 95, 101–2, 177. *See also* compensation; People and Property Count (PPC)
village land (category), 31–32
Village Land Act (1999), 31, 46, 83, 93, 153
villagization. See *ujamaa*: villagization
voluntary codes of conduct for investors and states, 184, 215n84

Wami River, *21–22*, 56, *57–59*, 59, *70*, 72. *See also* floodplain agriculture
Wami-Ruvu river basin, *22*, 54, *70*
Wami Sisal Estate, 73–75, 202n275
wanaharakati (activist group), 123–26, 131, 140–48, 177, 179

water: for plantations, 29, 35, 39, 71–72, 88, 213n45; for subsistence, 81, 84, 87, 99, 110, 131, 134, 180–81, 200n28; for wildlife, 84. *See also* flood events; floodplain agriculture; land: land-water as coupled resource; Wami River
weeds, 54, 60–61, 64–65
Weil, Simone, 184
White, Luise, 126
whiteness, 19–20, 194n96. *See also* race
widows, 7, 81, 95–96, 98–99, 103, 105, 126, 137–38, 156, 178, 181. *See also* women: "without men"
wildlife conservation, 4, 83–89, 128, 169, 175, 181–82
Wildlife Division, 84–85, 202n94
Williams, Raymond, 8
Winde (settlement), 53, 75, 77, 79–80, 136, *137*; lawsuit against RAZABA ranch, 80
Wise, Timothy, 11
Wolford, Wendy, 207n8
women: exclusion from political participation, 125, 145–47; gossip and, 129, 131–32; harassment and violence against, 117–18, 138–40; initiation rites, 55, 66, *67–68*, 68–69, 72, 95, 200n32 (see also *mkole*); as property, 14–15; "without men," 95, 98–103, 110, 145 (*see also* divorce; single mothers; widows); as wives and co-wives, 15, 74, 95–98, 110. *See also* female solidarity; femininity; gender
World Bank, 30, 41, 103, 203n1. *See also* International Centre for Settlement of Investment Disputes (ICSID)
World Economic Forum, 42–43
World Wildlife Fund (WWF), 30, 199n119

Yara International, 42, 43

Zanzibar: charcoal smuggling to, 135–36; Omani Sultanate of, 1, 53, 77; ranch in Bagamoyo (*see* RAZABA ranch)
Zaramo (ethnic group), 53, 59, 66, 200n32
Zigua (ethnic group), 59, 66, 157